Mechanisms and Machine Science

Volume 96

This book series establishes a well-defined forum for monographs, edited Books, and proceedings on mechanical engineering with particular emphasis on MMS (Mechanism and Machine Science). The final goal is the publication of research that shows the development of mechanical engineering and particularly MMS in all technical aspects, even in very recent assessments. Published works share an approach by which technical details and formulation are discussed, and discuss modern formalisms with the aim to circulate research and technical achievements for use in professional, research, academic, and teaching activities.

This technical approach is an essential characteristic of the series. By discussing technical details and formulations in terms of modern formalisms, the possibility is created not only to show technical developments but also to explain achievements for technical teaching and research activity today and for the future.

The book series is intended to collect technical views on developments of the broad field of MMS in a unique frame that can be seen in its totality as an Encyclopaedia of MMS but with the additional purpose of archiving and teaching MMS achievements. Therefore, the book series will be of use not only for researchers and teachers in Mechanical Engineering but also for professionals and students for their formation and future work.

The series is promoted under the auspices of International Federation for the Promotion of Mechanism and Machine Science (IFToMM).

Prospective authors and editors can contact Mr. Pierpaolo Riva (publishing editor, Springer) at: pierpaolo.riva@springer.com

Indexed by SCOPUS and Google Scholar.

More information about this series at http://www.springer.com/series/8779

Lena Zentner · Steffen Strehle
Editors

Microactuators, Microsensors and Micromechanisms

MAMM 2020

Springer

Editors
Lena Zentner
Department of Mechanical Engineering
Technische Universität Ilmenau
Ilmenau, Germany

Steffen Strehle
Department of Mechanical Engineering
Technische Universität Ilmenau
Ilmenau, Germany

ISSN 2211-0984 ISSN 2211-0992 (electronic)
Mechanisms and Machine Science
ISBN 978-3-030-61651-9 ISBN 978-3-030-61652-6 (eBook)
https://doi.org/10.1007/978-3-030-61652-6

This Springer imprint is published by the registered company Springer Nature Switzerland AG
The registered company address is: Gewerbestrasse 11, 6330 Cham, Switzerland

Preface

The 5th Conference on Microactuators, Microsensors and Micromechanisms (MAMM 2020) was held under the patronage of IFToMM (International Federation for the Promotion of Mechanism and Machine Science, http://www.iftomm.net). The goal of IFToMM is to promote research and development in the field of machines and mechanisms through theoretical and experimental methods and their practical application.

The MAMM conference was initiated by Professor Burkhard Corves in Aachen, where the first conference in 2010 took place as First Workshop on Microactuators and Micromechanisms. The second MAMM conference was held in Durgapur (India) in 2012 and the third in Timisoara (Romania) in 2014. After the fourth Conference in Ilmenau in 2016, the fifth conference MAMM returned to Ilmenau (Germany) in 2020 as the Conference on Microactuators, Microsensors and Micromechanisms (MAMM 2020) and took place from November 25–27, 2020.

The aim of the conference is to bring together scientists, industry experts and students from the field of miniaturized machines and mechanisms. The conference provides a special opportunity to intensify the exchange of knowledge and experience as well as to deepen the established collaborations in various disciplines of microactors, microsensors and micromechanisms. In the course of this event, presentations were made, which can be assigned to the following topics:

- Microactuators, microsensors and micromechanisms
- Compliant mechanisms and actuators
- Microscale power generation
- Miniaturized energy harvesting
- Micromanipulation, microassembly
- Miniature manufacturing of machines
- Mechatronics and control issues
- Micromechanical devices and robotics for life science
- Biomechatronics at small scale

Despite the challenging circumstances of year 2020, this book consists nonetheless of 12 peer-reviewed papers. The conference was characterized by international participation. Participants came from Germany, Serbia, Finland, USA and Peru.

Our special thanks to IFToMM for their support of the conference and of course to all authors and participants, who made the conference a success with their interesting contributions and inspiring scientific discussions. Last but not least, we would also like to thank the International Scientific Committee for its support, and we have always greatly appreciated the pleasant cooperation. The contributions of the 5th Conference on Microactuators, Microsensors and Micromechanisms 2020 were peer-reviewed by experts:

- Valter Böhm (Germany)
- Tzu-Chi Chan (Taiwan)
- Burkhard Corves (Germany)
- Antal Huba (Hungary)
- G. R. Jayanth (India)
- Robert Kirchner (Germany)
- Roy Knechtel (Germany)
- Julia Körner (Germany)
- Nenad T. Pavlovic (Serbia)
- Christoph Weigel (Germany)
- Hartmut Witte (Germany)
- Klaus Zimmermann (Germany)

Furthermore, we would like to thank the local organizing committee of the 5th MAMM 2020 at the Technische Universität Ilmenau for their outstanding support in organizing this scientific event and thus, for their assistance in all related activities. Special thanks to Alexandra Griebel, Stefan Henning, Maria Illing, Heidi Kirsten, Kirsti Schneider, Christoph Weigel and Dirk Wetzlich. The employees of Springer-Verlag, especially Mr. Pierpaolo Riva, have provided excellent technical and editorial support that we highly acknowledge. We hope you enjoy reading the articles and that they will give you new impulses and inspiration for further development and research in the fields of microactuators, microsensors and micromechanisms.

November 2020

Lena Zentner
Steffen Strehle

Contents

A Novel Planar Two-Axis Leaf-Type Notch Flexure Hinge with Coincident Rotation Axes and Its Application to Micropositioning Stages

Sebastian Linß[1](✉), Philipp Gräser[2], Mario Torres[2], Tobias Kaletsch[1], René Theska[2], and Lena Zentner[1]

[1] Compliant Systems Group, Technische Universität Ilmenau, 98693 Ilmenau, Germany
sebastian.linss@tu-ilmenau.de
[2] Precision Engineering Group, Technische Universität Ilmenau, 98693 Ilmenau, Germany

Abstract. Compliant mechanisms with flexure hinges are well-suited for high-precision applications due to their smooth and repeatable motion. However, the synthesis of planar compliant mechanisms based on notch flexure hinges is mostly limited to the use of single-axis hinges due to the lack of certain multiple-axis flexure hinges. This contribution introduces a novel planar leaf-type notch flexure hinge with two coincident rotation axes based on circular pre-curved leaf springs. A generally suitable hinge geometry is determined through a parametric study using the finite element method (FEM). Finally, the two-axis flexure hinge is applied and investigated for the use in two planar micropositioning stages for the rectilinear guidance of an output link with a large centimeter stroke. The presented two-axis flexure hinge turns out to be a suitable approach to monolithically connect three links of a compliant mechanism in a planar and precise way.

Keywords: Compliant mechanism · Flexure hinge · Two-axis leaf-type notch flexure hinge · Micropositioning stage · FEM analysis

1 Introduction

Due to their beneficial properties, compliant mechanisms show increasing potential for the application in many technical areas, especially in precision engineering and microtechnology [1]. The synthesis of compliant mechanisms for high-precision applications is mainly realized by replacing the rotational joints of a suitable rigid-body model with flexure hinges of standard notch shapes [2], e.g. with a semi-circular contour [3]. The absence of kinetic friction and wear combined with a well-defined rotation axis allows for a very precise and repeatable motion. A further improvement of the motion and deformation properties can be attained by using flexure hinges with optimized notch shapes based on power function contours [4]. Although there are numerous other flexure hinge types [5], the selection of a rigid-body model for planar precision applications is still very limited to mechanisms with only single-axis rotational joints in the kinematic

L. Zentner and S. Strehle (Eds.): MAMM 2020, MMS 96, pp. 1–14, 2021.
https://doi.org/10.1007/978-3-030-61652-6_1

chain. Thus, due to the lack of investigations on multiple-axis flexure hinges, only two links can be connected by existing flexure hinges in a planar way.

In the literature, two-axis rotational joints in compliant mechanisms for planar applications are usually approximated by two adjacent coplanar notch flexure hinges [1, 6–8] (see Fig. 1(a)), which connect two links to a third one. However, this concept does not possess a unique rotation axis for all the links, which may lead to a significant deviation of the expected in-plane motion in comparison to the rigid-body model. A further possibility is to use two non-planar single-axis flexure hinges with coincident rotation axes [9], as shown in Fig. 1(b). Still, a planar layout is not possible and, thus, it introduces undesired out-of-plane deformations and increases the manufacturing effort for a monolithic construction via wire EDM (electrical discharge machining). In [10], a three-link flexure hinge where all their axes intersect in the same point is applied in a vehicle seat. Although there is a unique ideal rotation axis, the design is not suitable for precision applications.

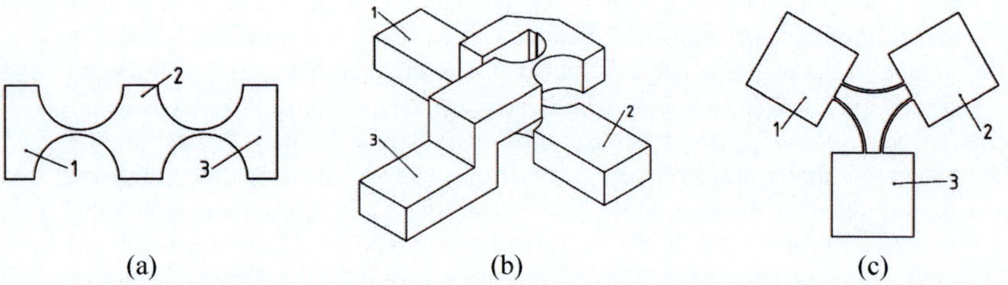

Fig. 1. Two-axis flexure hinges for monolithically connecting three links in planar compliant mechanisms: (a) adjacent coplanar notch hinges with parallel axes, (b) non-planar flexure hinge with coincident axes, (c) novel planar leaf-type notch flexure hinge with coincident axes.

In conclusion, the current state of the art does not include a monolithic planar two-axis or multiple-axis flexure hinge suitable for precision applications due to the aforementioned problems. Therefore, the present contribution introduces a flexure hinge with two coincident rotation axes based on a leaf-type notch flexure hinge with circular pre-curved segments (see Fig. 1(c)). The elasto-kinematic properties are investigated under the variation of the geometric parameters using accurate FEM models. Design guidelines and suitable parameter values are then determined for its use in high-precision compliant mechanisms. Finally, the potential of the novel planar two-axis flexure hinge is further proven by the application in two compliant micropositioning stages.

2 Design of the Two-Axis Leaf-Type Notch Flexure Hinge

The novel planar flexure hinge is, in principle, built by connecting the three links to one another using three pre-curved leaf springs. As a result, there is a compliant connection between each link, which differentiates the design with the state of the art. The curvature of the leaf springs can describe different geometries, similar to a notch flexure hinge. Therefore, it has been designated as two-axis leaf-type notch flexure hinge (TLNFH).

For the following investigations, a circular geometry is selected for the leaf design. Due to existing investigations on similar single-axis notch flexure hinges with circular pre-curved segments [11–13], a better comparability of results is possible. Also, only fully symmetrical geometries around the longitudinal and transversal axes are considered to limit the complexity of the flexure hinges. Specific and non-intuitive optimization approaches for multi-notched hinges, like suggested by [14], are not regarded here, as they cannot be generalized for the use in whole mechanisms.

For a general description of the novel flexure hinge design, a universally applicable selection of parameters is necessary. The parameters are oriented on the definition for a single hinge given by [15], see Fig. 2. L is the total hinge length and l is the length of the notch region for the case of a single-axis flexure hinge. H describes the total height of the links, w is the width, and β_1, β_2 are the angles between the axes of the links. Particularly relevant for the following considerations are the three investigated main design parameters, the leaf thickness t, the minimum gap between two leaf springs of the same link h, and the transition angle between a link and a leaf spring γ. Due to constructive and manufacturing aspects, a fillet of radius r is required as the transition between the leaf springs and the links. To reduce the number of studied parameters, the value of each parameter is the same in every element of the flexure hinge.

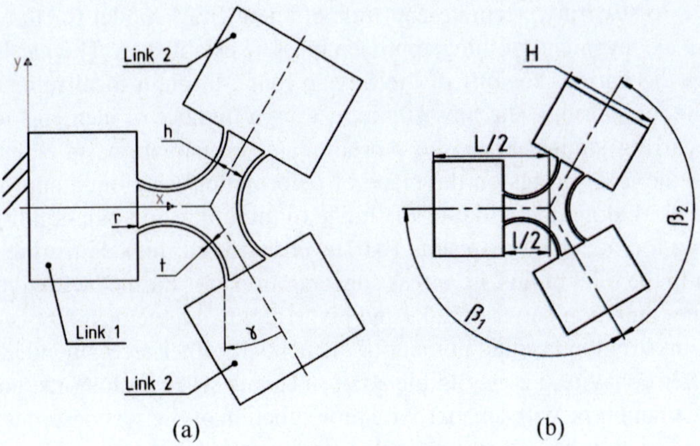

Fig. 2. Geometric parameters of the TLNFH: (a) hinge model with main variation parameters γ, t and h, (b) hinge parameters with constant values for this investigation.

3 FEM-Based Investigation of the TLNFH

In this section, the influence of three main geometric parameters of TLNFH are investigated using a quasi-static structural 3D FEM simulation model to evaluate the properties of the novel design, specifically the maximum equivalent strain, the rotation axis shift and the input force. The geometric parameters in Table 1 are assumed as the starting point for the parametric study of the transition angle γ, the leaf thickness t, and the minimum leaf gap h.

Table 1. Overview of the geometric parameters of the TLNFH with basic values.

Parameter	Symbol	Value	Parameter	Symbol	Value
Leaf thickness	t	0.3 mm	Hinge width	w	6 mm
Minimum leaf gap	h	0.3 mm	Total hinge height	H	10 mm
Transition angle	γ	0°	Total hinge length	L	50 mm
Hinge length	l	10 mm	Relative link angle 1	β_1	120°
Fillet radius	r	0.1 mm	Relative link angle 2	β_2	120°

Based on the obtained results, the suitable values of these parameters are determined for the application in example mechanisms. A typical hard aluminum alloy EN AW 7075 with a Young's modulus of $E = 72$ GPa, a Poisson's ratio of $\nu = 0.33$ and maximum admissible elastic strain of $\varepsilon_{\text{adm}} = 0.5\%$ is selected as the material for the FEM-based investigations using the commercial software ANSYS Workbench 18.2.

3.1 FEM Model

To generate a trustworthy, accurate and time-efficient FEM model for the parametric investigation, a convenient meshing approach must be established. The mesh is defined differently in the various regions of the flexure hinge through local refinements (see Fig. 3). In the leaf springs, the mesh requires a high number of elements to correctly seize the occurring strains. Based on a preliminary investigation, an element size of 0.075 mm in the leaf springs on the plane of deformation was found appropriate. Six elements are used along the width of the hinge to minimize computation time without compromising the accuracy of results (0.01% error for strains). Since the maximum strains occur in the fillet radius, i.e. stress concentration, it is meshed with five elements. To smooth the transition between links and hinge region, an element size of 0.5 mm is used. In the links region, a small number of elements is rather recommended due to the expected stiffer behavior, i.e. negligible strains. This is set to 1 along its length. A 3D hexahedral element type with an approximation function of the second-order is selected for the mesh. The simulations are conducted considering geometric nonlinearities.

Due to the higher number of links in comparison to a typical flexure hinge, multiple load cases can be studied. An important aspect of the investigation is the comparability of the cases with the motion when using the hinge in a mechanism. The load cases and corresponding boundary conditions must depict the behavior in the most realistic way possible. Thus, the following boundary conditions are set (cf. Figure 2):

- Case 1: Link 1 fixed, links 2 and 3 rotate counterclockwise (or clockwise),
- Case 2: Link 1 fixed, link 2 rotates clockwise and link 3 rotates counterclockwise,
- Case 3: Link 1 fixed, link 2 rotates counterclockwise and link 3 rotates clockwise.

The motion direction for load case 1 can be indistinctively clockwise or counter-clockwise due to the symmetry of the TLNFH. The rotation of links 2 and 3 in load

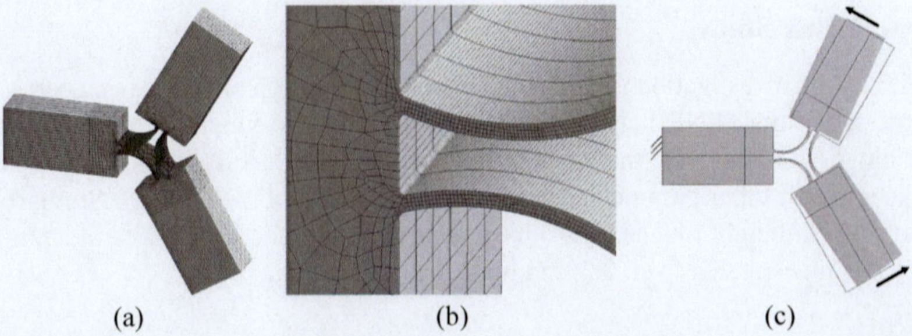

Fig. 3. 3D FEM model of the TLNFH: (a) overall mesh, (b) detail of the leaf mesh, (c) deformed state of the hinge with boundary conditions for load case 1 (two transverse force loads).

cases 2 and 3 is symmetric about the *x*-axis. For the investigations, a discrete rotation of $\varphi = 5°$ of links 2 and 3 is realized iteratively through a direction-constant transverse force load applied in several load steps.

To determine the axis shifts of the TLNFH, a modeling approach is established due to the absence of an ideal rotation axis of the three links. Here, the intersection of the axes of the links is assumed as the ideal rotation axis. The axis shift is defined as the distance traveled by the guided fixed intersection point from its initial position to its position after deformation, i.e. fixed center approach [15]. Thus, in contrast to a single-axis flexure hinge, more than one axis shift can be modeled by the relative motion of two links in the TLNFH. The three axis shifts ν_{ij} present in the TLNFH are represented by arrows in Fig. 4. Due to the symmetry of the geometry and load cases, the values of each ν_{ij} lie around the same order of magnitude. As such, only the relative deviation ν_{21} is evaluated in the following investigation and is designated as ν_D from here onwards.

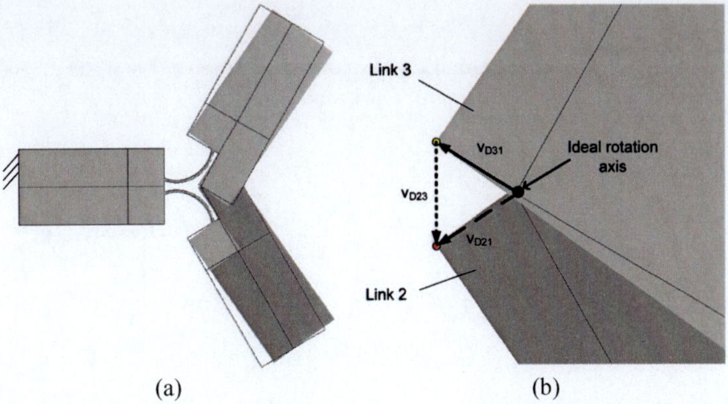

Fig. 4. Model-based determination of the rotational axis shifts of the TLNFH for load case 2: (a) FEM model with deformation of the hinge, (b) detailed definition of the resulting axis shifts.

3.2 Parametric Study

The FEM-based investigation is conducted under the variation of the relevant parameters for the design of the TLNFH and the evaluation of the elasto-kinematic properties of the flexure hinge. In a first step, the value of the transition angle γ is varied from 0° to 90° while keeping all other parameters constant (see Fig. 5 and cf. Table 1). Similarly, the value of the leaf height t is varied from 0.3 mm to 1 mm (see Fig. 6) and the minimum leaf gap h is investigated from 0.3 mm up to 4 mm (see Fig. 7).

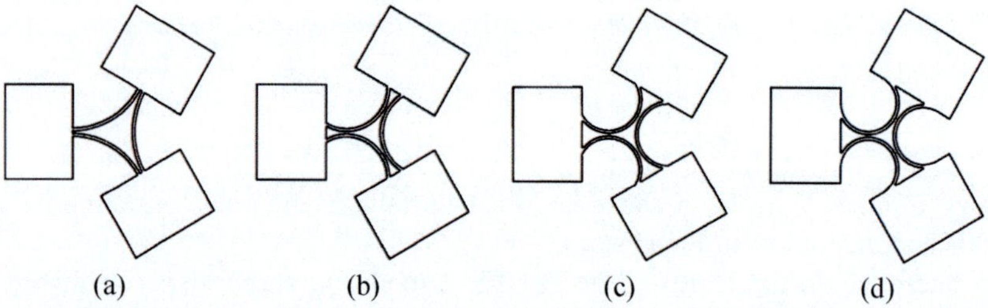

Fig. 5. Flexure hinge examples for variation of transition angle γ for $h = 0.3$ mm and $t = 0.3$ mm: (a) $\gamma = 0°$, (b) $\gamma = 30°$, (c) $\gamma = 60°$, (d) $\gamma = 90°$.

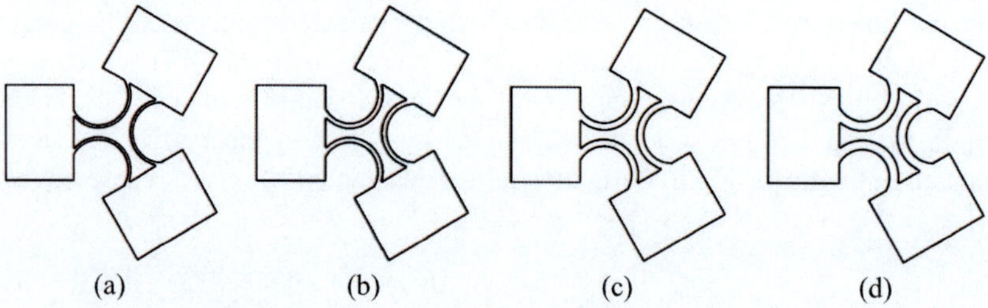

Fig. 6. Flexure hinge examples for variation of leaf thickness t for $h = 1$ mm and $\gamma = 60°$: (a) $t =$ 0.3 mm, (b) $t = 0.5$ mm, (c) $t = 0.7$ mm, (d) $t = 1$ mm.

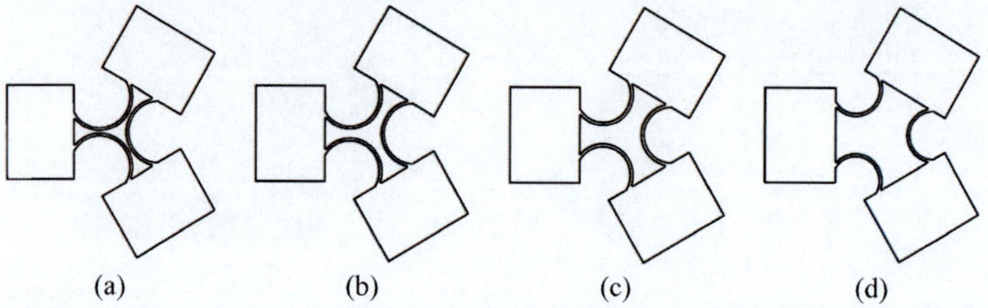

Fig. 7. Flexure hinge examples for variation of minimum leaf gap h for $t = 0.3$ mm and $\gamma = 60°$: (a) $h = 0.3$ mm, (b) $h = 1$ mm, (c) $h = 2$ mm, (d) $h = 4$ mm.

The results of the maximum von Mises equivalent strain $\varepsilon_{\mathrm{v,max}}$, rotational axis shift ν_{D} and input force F_{21} (for link 2 as example) are presented in Fig. 8, 9 and 10.

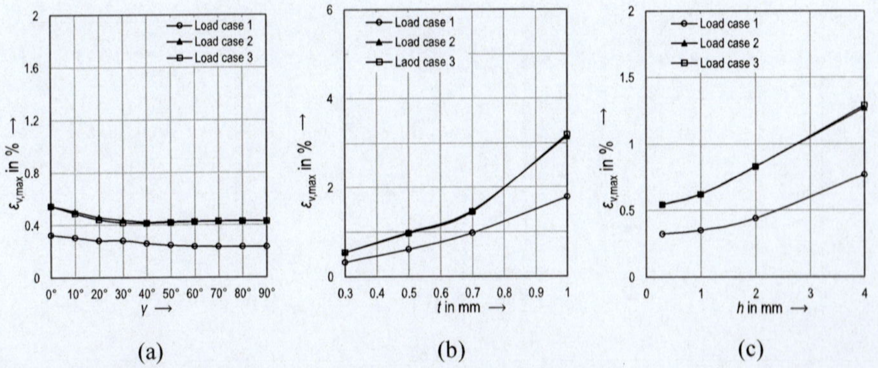

Fig. 8. FEM results of the maximum von Mises equivalent strain $\varepsilon_{v,max}$ relative to: (a) transition angle γ, (b) leaf thickness t, (c) minimum leaf gap h.

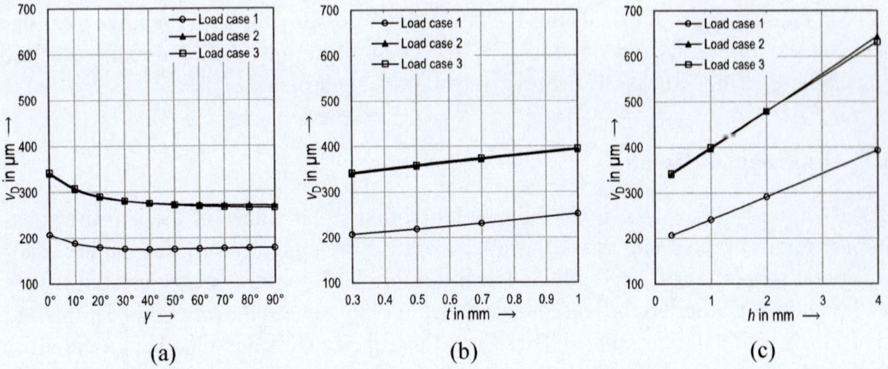

Fig. 9. FEM results of the rotational axis shift v_D relative to: (a) transition angle γ, (b) leaf thickness t, (c) minimum leaf gap h.

Regarding $\varepsilon_{v,max}$, small values of γ produce expectedly higher strains (see Fig. 8(a)) due to the abrupt change of the cross-section from the hinge to the link region, i.e. stress concentration. Above $\gamma = 30°$ to $40°$ the strains reach an almost constant value, possibly due to the maximum values occurring around the regions where the gap between the leaf springs is at its minimum. As shown in Fig. 9(b), an increasing leaf thickness leads to higher strains. This is mainly due to the increased stiffness and, thus, the required input forces to rotate each neighbored link with $\varphi = 5°$. The minimum leaf gap follows a similar trend to the leaf thickness (see Fig. 8(c)). As h increases, $\varepsilon_{v,max}$ increases accordingly. Although the geometry of the leaf springs has not been altered, the moment of inertia of the cross-section is increased with h. This leads to higher stresses and strains in the leaf springs.

The results of the axis shift and input force follow a similar trend to the strain results (see Fig. 9 and 10), since there is a close relationship between them. As the strain increases, the deformation force and, thus, the axis shift becomes larger. It is also to remark that the values of $\varepsilon_{v,max}$, v_D and F_{21} are consistently higher in the load cases 2

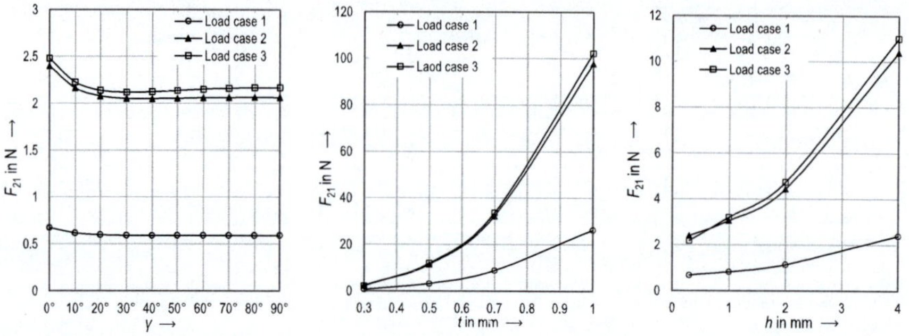

(a) (b) (c)

Fig. 10. FEM results of the input force F_{21} for link 2 relative to: (a) transition angle γ, (b) leaf thickness t, (c) minimum leaf gap h.

and 3 in comparison to the load case 1. The reason is possibly the larger relative rotations between the links 2 and 3 around $|\varphi_{\mathrm{rel}}| = 10°$ due to different rotation directions of the links, whereas, in load case 1, the relative rotations lie around $|\varphi_{\mathrm{rel}}| = 5°$.

3.3 Discussion of Results

Based on the FEM results, it can be concluded that small values of the leaf thickness and the minimum leaf gap as well as a large transition angle produce favorable elasto-kinematic properties of the TLNFH. The minimum leaf gap has the strongest influence on the axis shift, whereas the leaf thickness strongly influences the maximum strains and input forces of the flexure hinge. Therefore, these parameters are to be kept as small as possible in the design of TLNFH. However, the achievable minimum values are typically limited to $t = 0.3$ mm and $h = 0.3$ mm due to manufacturing limitations, i.e. EDM with standard wire. The influence of geometric tolerances due to non-ideal manufacturing is mostly determined by the investigated variation of t. The use of smaller leaf gaps could also lead to contact between the leafs, causing a nonlinear and non-repeatable behavior of the hinge, like described by [13]. A further possibility for the minimum leaf gap is the union between the leafs, i.e. $h \leq 0$. Still, such a flexure hinge is very related to the single-axis ones investigated in [14], where there is no significant influence on the rotational precision or the stress/strain behavior. In contrast, the transition angle requires to be maximized up to 90° for optimal properties of the TLNFH.

Using the obtained knowledge, a comparison between a TLNFH with unfavorable and one with suitable parameter values is presented in Table 2. A suitable geometry of the TLNFH can reduce the maximum strain up to 14 times and the axis shift up to 2.5 times. The TLNFH shows a linear deformation and motion behavior in general.

Table 2. Resulting elasto-kinematic properties of the TLNFH with two opposite designs.

		Unfavorable design of the TLNFH	Suitable design of the TLNFH
		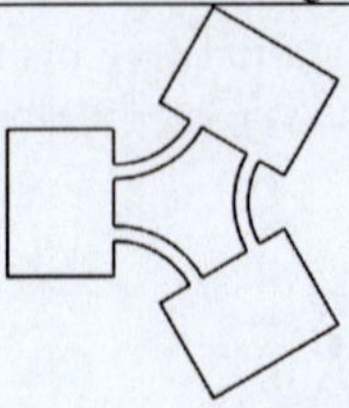 $\gamma = 0°$, $h = 1$ mm, $t = 1$ mm	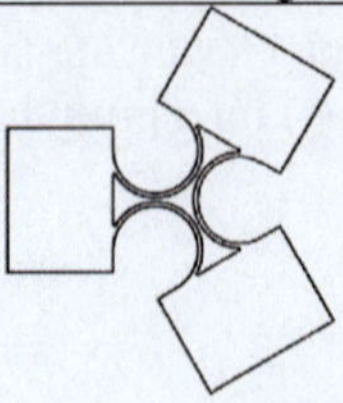 $\gamma = 90°$, $h = 0.3$ mm, $t = 0.3$ mm
Load case 1	$\varepsilon_{v,max}$	3.48 %	0.2 %
	v_D	451 µm	177 µm
	F_{21}	96.2 N	0.6 N
	F_{31}	84.8 N	0.6 N
Load case 2	$\varepsilon_{v,max}$	5.41 %	0.43 %
	v_D	674 µm	270 µm
	F_{21}	391.5 N	2.1 N
	F_{31}	391.0 N	2.1 N
Load case 3	$\varepsilon_{v,max}$	5.42 %	0.2 %
	v_D	663 µm	265 µm
	F_{21}	417.3 N	2.2 N
	F_{31}	417.1 N	2.2 N

4 Application to Compliant Micropositioning Stages

For the verification of the potential of the novel TLNFH with regard to planar mechanisms, two micropositioning stages based on monolithic planar compliant mechanisms are used as application examples. The first one concerns a variation of the well-known ROBERTS mechanism for producing an approximated straight-line motion of the output link. The second one is an enhanced pantograph mechanism with two orthogonal motion axes. For each mechanism, the rigid-body model is compared with its realization as compliant mechanism using the novel TLNFH for the two-axis joints. According to the synthesis method presented in [4], the remaining single-axis joints are replaced by suitable notch flexure hinges with a power-function contour. For this, an exponent equal to 4 for the ROBERTS mechanism and to 5 or 6 for the pantograph mechanism is used to not exceed the admissible strain in dependence of the hinge rotation angles, which are always smaller than 7.5°. The mechanisms are dimensioned here to realize a motion range of ±10 mm along each axis. The investigation is done using FEM models.

4.1 10-Hinge Rectilinear Stage

The investigated 10-hinge mechanism after ROBERTS is shown in Fig. 11. It possesses eight links with two two-axis and six one-axis rotational joints. The used hinge lengths are $\overline{A_0A} = \overline{B_0B_1} = \overline{C_0C_1} = 87.04$ mm, $\overline{AB_1} = 88.4$ mm, $\overline{B_0C_0} = \overline{BC} = \overline{ED} = 20.0$ mm, $\overline{A_0B_0} = 149.9$ mm and $\overline{CD} = 171.50$ mm using the replacement position with α of 69.63°. Thus, the rigid-body model realizes an approximated rectilinear plane guidance motion of the coupler 8 with a small straight-line deviation Δy_K in the micrometer range (30 μm for point K) for a given input displacement Δx_K of ± 10 mm.

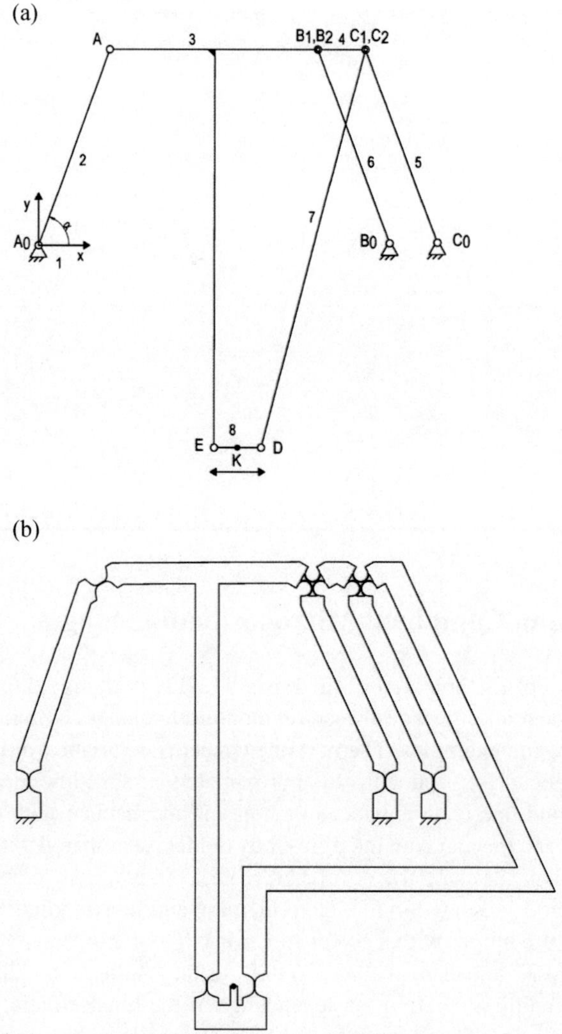

Fig. 11. 10-hinge ROBERTS mechanism: (a) rigid-body model, (b) realization as compliant mechanism using six single-axis power function notch flexure hinges and two TLNFH.

According to the FEM simulations (Table 3), the compliant mechanism using the TLNFH and six single-axis power function flexure hinges shows a very small straight-line deviation (<5 μm) when the deflection is negative. For a positive deflection, its behavior is very similar to the rigid-body model as represented by the comparing path deviation p (<6 μm). In general, the results have shown a good agreement between the compliant and the rigid-body mechanism with a deviation of less than 30 μm while keeping the maximum strain below the admissible value of 0.5%.

Table 3. Elasto-kinematic properties of the compliant 10-hinge ROBERTS mechanism.

Δx_K in mm	Δy_K in μm	p in μm	$\varepsilon_{v,max}$ in %	Input force in N
−10	−3.59	25.84	0.45	2.91
−8	0.11	18.99	0.36	2.32
−6	2.36	12.99	0.27	1.74
−4	3.11	7.88	0.18	1.16
−2	2.33	3.52	0.09	0.58
2	−3.86	−2.69	0.09	0.57
4	−9.26	−4.57	0.18	1.15
6	−16.21	−5.64	0.27	1.72
8	−24.65	−5.85	0.37	2.28
10	−34.55	−5.26	0.46	2.85

4.2 12-Hinge XY Stage

The investigated and enhanced 12-hinge pantograph mechanism with two degrees of freedom is designed after [8]. The mechanism allows two orthogonal rectilinear motions of the output link 11, as shown in Fig. 12. It possesses eleven links with three two-axis and six one-axis rotational joints as well as two prismatic joints. The used hinge lengths are $\overline{AB} = \overline{CD} = \overline{EF} = 28.28$ mm, $\overline{DE} = \overline{CF} = \overline{BG} = \overline{FH} = \overline{GI} = 113.14$ mm, $\overline{BC} = \overline{GF} = \overline{IH} = 28.0$ mm, $\overline{AB_x} = 40$ mm and $\overline{AB_y} = 28.0$ mm. The input links 2 and 4 are connected to the frame, so that due to the parallel arrangement none of the drive elements is guided. Also, the two output axes are each assigned to a drive decoupled from the other in the rigid-body model. An input displacement of link 2 produces a motion of the coupler point K only in x-direction with a transmission ratio of −0.25. Analogously, the link 4 defines the position of K only in y-direction with a ratio of 0.2.

The two prismatic joints for the linear x and y input guidance motions are simplified as direction-constant displacements of the input links 2 and 4 in the FEM simulation to focus on modeling the impact of the novel TLNFH. Thus, according to [1, 5] the investigated 12-hinge pantograph mechanism represents a partially compliant mechanism.

Still, a positioning deviation occurs in the realization as compliant mechanism as well as a slight rotation of the output link. The deviations in the x- and y-direction according to the FEM results are depicted in Fig. 13. A maximum deviation in the sub-millimeter range is produced due to the axis shift of all the flexure hinges. However, the rather linear and highly reproducible behavior of the deviation allows a simple compensation

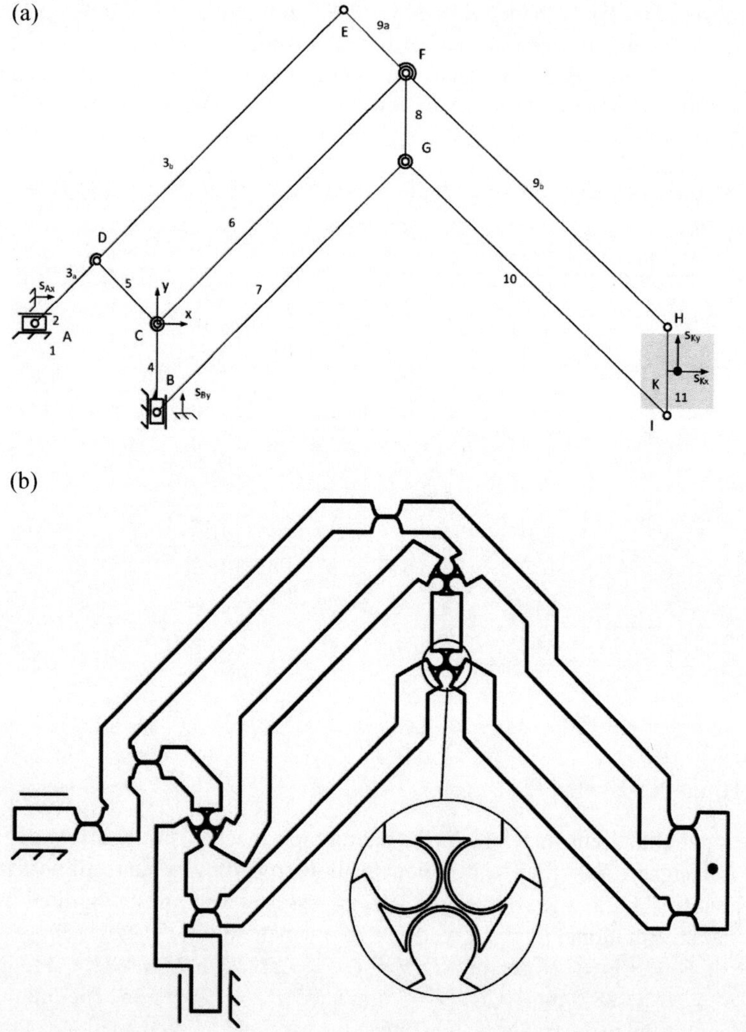

Fig. 12. 12-hinge pantograph mechanism: (a) rigid-body model, (b) realization as compliant mechanism using six single-axis power function notch flexure hinges and three TLNFH.

to increase the motion precision. Also, the use of optimized contours for the two-axis flexure hinges may further improve the motion behavior of the mechanism.

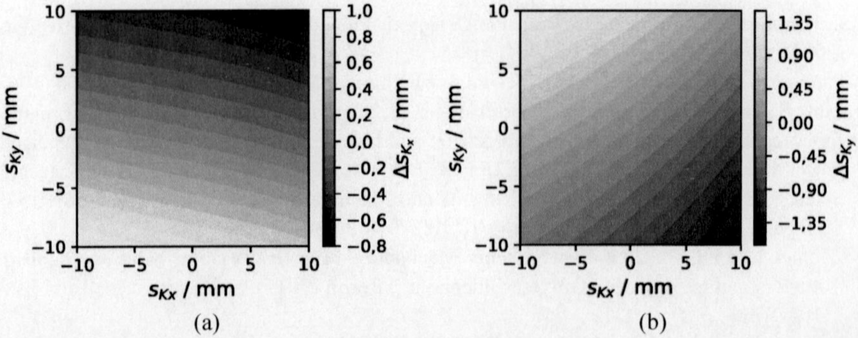

Fig. 13. Positioning deviations of the coupler point K for the 12-hinge pantograph mechanism relative to the coupler point displacements s_{Kx} and s_{Ky}: (a) in x-direction, (b) in y-direction.

5 Conclusions

The synthesis of a planar compliant mechanism based on the rigid-body model is currently limited to the use of only single-axis flexure hinges. An existing approach for designing two-axis flexure hinges with coincident rotation axes is the use of non-planar layouts. However, due to the increased complexity, a monolithic design, which is a major advantage of compliant mechanisms, is either technically not possible or requires a significant manufacturing effort. The novel TLNFH introduced in this contribution exhibits enormous potential for the application in higher-bar planar compliant mechanisms. The presented investigations on the two-axis flexure hinge and on the application in mechanisms have shown that the TLNFH is well suited for this purpose. According to the FEM simulations, the resulting positioning deviations of a rectilinear guiding mechanism using the TLNFH evidence a highly precise motion. A motion behavior can be attained here in some ranges of deflection, which deviates from the rigid-body model by only a couple of micrometers. For a mechanism with two degrees of freedom, the positioning deviations reach, however, a higher degree. Due to the highly reproducible motion and linear behavior of the deviation, a compensation is possible. The use of optimized power-function-based contours can also be used to further improve the motion behavior of this new type of flexure hinge. Thus, the novel planar design of a two-axis flexure hinge with coincident rotation axes brings about new possibilities to the design of planar compliant mechanisms for precision applications. The presented two-axis flexure hinge can be beneficially used in any multi-bar compliant mechanism to connect three links in a monolithic, planar and precise way.

Acknowledgements. The authors would like to gratefully acknowledge the support of the German Research Foundation (DFG) under grant numbers ZE 714/10-2 and TH 845/5-2.

References

1. Howell, L.L., Magleby, S.P., Olsen, B.M.: Handbook of Compliant Mechanisms. Wiley, Chichester (2013)

2. Lobontiu, N.: Compliant Mechanisms: Design of Flexure Hinges. CRC Press, Boca Raton (2003)
3. Paros, J.M., Weisbord, L.: How to design flexure hinges. Mach. Des. **25**(11), 151–156 (1965)
4. Linß, S., Gräser, P., Henning, S., Harfensteller, F., Theska, R., Zentner, L.: Synthesis method for compliant mechanisms of high-precision and large-stroke by use of individually shaped power function flexure hinges. In: Uhl, T. (eds.) Advances in Mechanism and Machine Science. IFToMM WC 2019. Mechanisms and Machine Science, vol. 73, pp. 1569–1578. Springer, Cham (2019). https://doi.org/10.1007/978-3-030-20131-9_155
5. Zentner, L., Linß, S.: Compliant Systems: Mechanics of Elastically Deformable Mechanisms, Actuators and Sensors. De Gruyter Oldenbourg, Berlin (2019). https://doi.org/10.1515/9783110479744
6. Holzem, H.: Gelenk. German Patent DE4030157C2 (1992)
7. Noll, T., Kalus, C., Peatman, W.B., Schade, U., Scheibner, W.V.: The monolithic two axis flexure joined mirror support and the mechanical design of the infrared beamline. In: 2nd International Workshop on Mechanical Engineering Design of Synchrotron Radiation Equipment and Instrumentation (MEDSI02), Illinois, pp. 65–73 (2002)
8. Horie, M., Hoshikawa, Y., Kamiya, D.: Optimum design of mass distribution of the injection molding pantograph mechanism with constant output link orientation. In: Ananthasuresh, G.K., Corves, B., Petuya, V. (eds.) Micromechanics and Microactuators 1st MAMM, Mechanisms and Machine Science, vol. 2, pp. 31–39. Springer, Dordrecht (2011). https://doi.org/10.1007/978-94-007-2721-2_4
9. Linß, S., Milojević, A., Zentner, L.: On the influence of flexure hinge geometry on the motion range and precision of compliant gripping mechanisms. In: Milcić, D. (ed.) Proceedings of the 2nd International Conference Mechanical Engineering in XXI Century (MASING 2013), pp. 255–260. Nis, Serbia (2011)
10. Schnapp, C., Amstein, M., Henning, S., Meyer, E., Erbacher, B., Hofmann, J.: Fahrzeugsitz. German Patent Application DE102015201232A1 (2016)
11. Yu, J., Zong, G., Yu, Z., Bi, S.: A new family of large-displacement flexural pivots. In: Proceedings of the ASME 2007 International Design Engineering Technical Conferences and Computers and Information in Engineering Conference, vol. 8, pp. 93–100. ASME (2007). https://doi.org/10.1115/detc2007-34977
12. Ahuett-Garza, H., Chaides, O., Garcia, P.N., Urbina, P.: Studies about the use of semicircular beams as hinges in large deflection planar compliant mechanisms. Precision Eng. **38**(4), 711–727 (2014). https://doi.org/10.1016/j.precisioneng.2014.03.008
13. Jin, M., Zhu, B., Mo, J., Zhou, Y., Zhang, X., Howell, L.L.: A CPRBM-based method for large-deflection analysis of contact-aided compliant mechanisms considering beam-to-beam contacts. Mech. Mach. Theory **145**, 103700 (2020). https://doi.org/10.1016/j.mechmachtheory.2019.103700
14. Liu, M., Zhang, X., Fatikow, S.: Design and analysis of a multi-notched flexure hinge for compliant mechanisms. Precision Eng. **48**(4), 292–304 (2017). https://doi.org/10.1016/j.precisioneng.2016.12.012
15. Linß, S., Schorr, P., Zentner, L.: General design equations for the rotational stiffness, maximal angular deflection and rotational precision of various flexure hinges. Mech. Sci. **8**(1), 29–49 (2017). https://doi.org/10.5194/ms-8-29-2017

Characterization of Thin Flexure Hinges for Precision Applications Based on First Eigenfrequency

Maximilian Darnieder[1(✉)], Felix Harfensteller[2], Philipp Schorr[3], Moritz Scharff[3], Sebastian Linß[4], and René Theska[1]

[1] Precision Engineering Group, Department of Mechanical Engineering, Max-Planck-Ring 12, 98693 Ilmenau, Germany
maximilian.darnieder@tu-ilmenau.de
[2] Engineering Design Group, Department of Mechanical Engineering, Max-Planck-Ring 12, 98693 Ilmenau, Germany
[3] Technical Mechanics Group, Department of Mechanical Engineering, Max-Planck-Ring 12, 98693 Ilmenau, Germany
[4] Compliant Systems Group, Department of Mechanical Engineering, Max-Planck-Ring 12, 98693 Ilmenau, Germany

Abstract. Flexure hinges with small cross-section heights are state of the art in numerous precision engineering applications due to their capability for smooth and repeatable motion. However, the high sensitivity to manufacturing influences represents a challenge. We propose a characterization method for flexure hinges based on the measurement of the free oscillation, to enable the consideration of manufacturing influences in the early stages of the design process. Three semi-circular flexure hinges with different cross-section heights and highly accurate geometry were investigated experimentally to compare them with three theoretical modeling approaches. The results for the three flexure hinge specimens showed small deviations to the predicted values from the models which is in agreement with the results of dimensional measurements. With each modeling approach, a deviation of the minimal notch height from the nominal value can be calculated. This value, in turn, can be used as manufacturing allowance for subsequent manufacturing of compliant mechanisms using the same manufacturing method. An exemplary compliant parallel-crank mechanism proves the applicability of the concept to compliant mechanisms with multiple flexure hinges.

Keywords: Compliant mechanism · Flexure hinge · Modal analysis · Manufacturing influences · Measurement

1 Introduction

Flexure hinge-based compliant mechanisms have many advantages over their counterparts with conventional joints and allow for a smooth and repeatable motion [1]. Hence,

L. Zentner and S. Strehle (Eds.): MAMM 2020, MMS 96, pp. 15–24, 2021.
https://doi.org/10.1007/978-3-030-61652-6_2

they are state of the art in numerous precision engineering applications [2]. For high precision force measurement tasks in metrology, mostly notch flexure hinges [3] with semi-circular shapes [4] are used. These flexure hinges are often designed with a minimal cross-section height in the micrometer range [5], which results in high sensitivities to manufacturing influences. The manufacturing process influences surface topography and the surface material properties, which are summarized under the term surface integrity [6].

As compliant mechanisms with thin flexure hinges are sensitive to vibrations, several modeling approaches in literature address the dynamic behavior of flexure hinges and compliant mechanisms, e.g. [7–9]. Ling et al. [10] provide a comprehensive overview and comparison of different modeling approaches. Furthermore, the accuracy of design equations, that describe the static properties based on analytical considerations [4, 11], strongly depend on the flexure hinge aspect ratios leading to different bending states, varying between plane stress or plane strain [12].

The design process of compliant mechanisms is commonly based on ideal geometries [13]. Depending on the manufacturing technology and the flexure hinge geometry, this assumption frequently results in pronounced deviations between the predicted and the measured values. Hence, it is desirable to quantify the manufacturing deviations of the used manufacturing process at the outset of the design process for the compliant mechanism.

We propose to manufacture and test single flexure hinge specimens in advance of or during the design process of a compliant mechanism. The measurement of the free oscillation and the extraction of the first eigenfrequency provides a highly effective measure for the characterization of the actual manufactured flexure hinge concerning the bending mode about its main bending axis. Among those are the rotational stiffness defined predominantly by the minimal notch height and the moment of inertia of the moving part. Using a model-based approach, the deviations from the ideal properties of the flexure hinge can be determined and used to adapt the mechanism design to the present manufacturing influences. This method is the focus of this paper, as well as the applicability and accuracy of corresponding modeling approaches in comparison with experimental results.

2 Materials and Methods

Three precision manufactured flexure hinges and a parallel-crank mechanism were analyzed both theoretically using three models and experimentally by measuring the free oscillation motion. The flexure hinges are manufactured by wire-EDM with highest diligence to reduce the geometrical deviations as much as possible. Since this technology is not always available, we aim at a combination of modeling and measurement to achieve equivalent results with more cost- and time-efficient manufacturing processes.

2.1 Experiment

The experimental foundation of this method is the contactless measurement of the free oscillation of flexure hinge specimens with semi-circular contour. Their first eigenfrequency comprises the mechanical properties of the flexure hinge for rotations about its principal bending axis. The eigenfrequency was determined using the setup in Fig. 1.

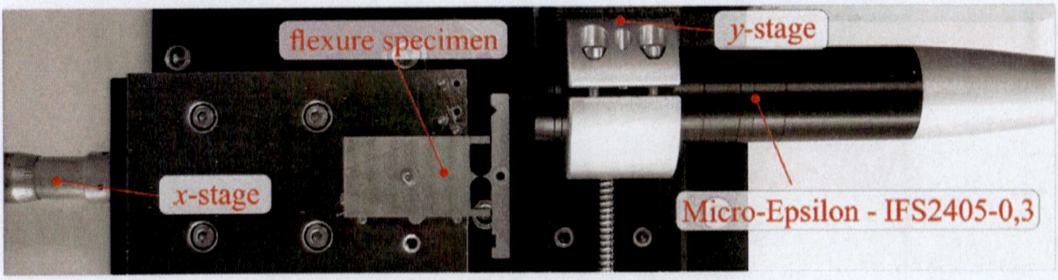

Fig. 1. Top view of the setup for the oscillation measurement with the flexure hinge specimen.

On the left-hand side the flexure hinge specimen is supported by a manually operated linear axis in x-direction to fine-position the flexure hinge in the measurement range of the sensor. The sensor is fixed to the carriage of a motorized axis oriented in y-direction. The sensor properties are summarized in Table 1. During the measurement sequence, the servo axis set different lateral positions relative to the flexure hinge. At each position, the oscillation measurement was repeated to identify random measurement errors resulting from local surface conditions. The measurement was repeated eleven times and the mean as well as the standard deviation were evaluated. Each time, the signal was recorded for 100 s. This results in a frequency resolution of the fast Fourier transform (FFT) of 0.01 Hz. The standard deviation was neglected since it was below the frequency resolution.

Table 1. Parameters of the experimental setup with displacement sensor (IF2405-0,3).

Parameter	Value	Unit
Sensor resolution	10	nm
Numerical aperture	0.6	
Spot diameter	6	µm
Linearity	<±15	µm
Measurement range	0.3	mm
Maximum permissible tilt angle	±34	°
Sampling rate	10	kHz

The flexure specimens were clamped with the local z-direction oriented in parallel to the vector of gravitational acceleration ($\vec{g}$). This horizontal orientation excludes the additional mechanical effect on the flexure hinge eigenfrequency, well known from the

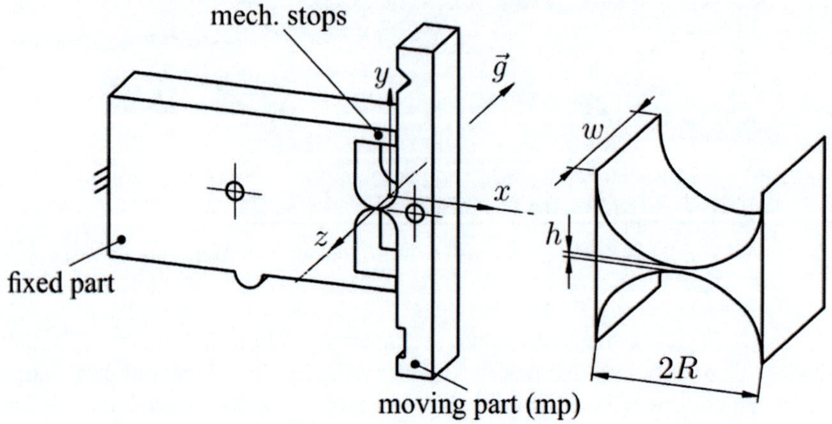

Fig. 2. Flexure hinge specimen on the left with its geometrical parameters on the right.

inverted pendulum. Three different flexure hinge specimens were analyzed. They are described in Fig. 2 and the properties are given in Table 2.

Table 2. Parameters of the flexure hinge specimens (EN AW-7075 T6).

Parameter	Symbol	Value	Unit
Minimal notch height	h	50; 75; 100	µm
Width	w	10	mm
Radius	R	3	mm
Yield strength	$R_{p0.2}$	503	MPa
Youngs modulus	E	71.7	GPa
Poisson number	ν	0.33	
Density	ρ	2800	kg m^{-3}
Mass moving part	m_{mp}	6.058	g
Moment of inertia (center of mass)	J_{mp}	999.985	g mm^2

Next to the three flexure hinges, a compliant parallel-crank mechanism, was analyzed using the same experimental setup. The parallel-crank mechanism is composed of four semi-circular flexure hinges as shown in Fig. 3. The geometrical and material parameters of the parallel crank are given in Table 3.

The experimental results were compared to three different theoretical approaches.

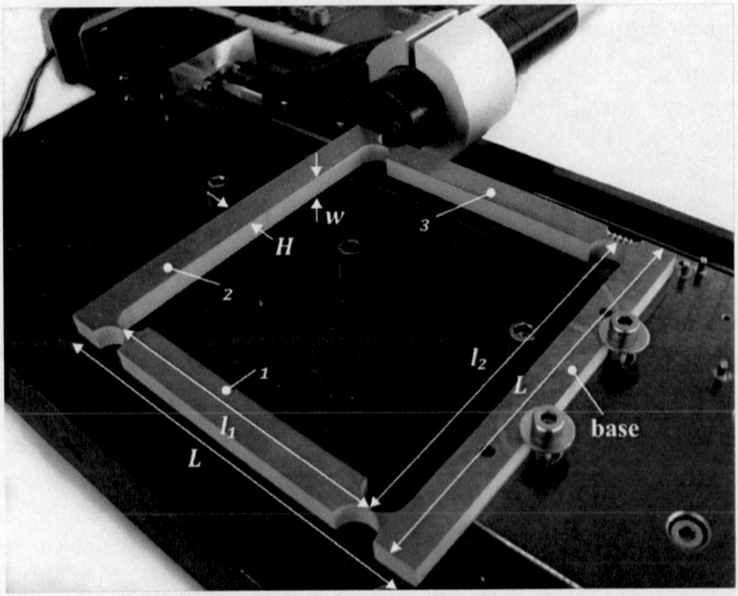

Fig. 3. Parallel-crank mechanism clamped in the experimental setup for the free oscillation measurement. The compliant mechanism has four semi-circular flexure hinges.

Table 3. Parameters of the parallel-crank mechanism (EN AW-7075 T6), compare Fig. 3.

Parameter	Symbol	Value	Unit
Minimal notch height	h	300	µm
Height	H	10	mm
Depth	w	6	mm
Radius	R	15	mm
Total length	L	110	mm
Member length	l_1, l_3	80	mm
	l_2	100	mm
Yield strength	$R_{p0.2}$	503	MPa
Young's modulus	E	71.7	GPa
Poisson number	ν	0.33	
Density	ρ	2800	kg m^{-3}
Mass	m_1, m_3	12.0	g
	m_2	19.0	g
Moment of inertia (center of mass)	J_1, J_3	5444	g mm^2
	J_2	19833	g mm^2

2.2 Theoretical Approaches

All three approaches are constrained by different assumptions and restrictions e.g. a neglected energy dissipation. The comparison between model and experiment is supposed to highlight whether the corresponding model captures all effects in a sufficient way.

Pseudo-Rigid-Body Model (PRBM)
The PRBM essentially follows the concept reviewed in [10]. We replaced the actual flexure hinge by a pivot and a torsional spring at the center, the moving and the fixed part by rigid bodies. The dynamics of the single flexure hinge was derived using the Lagrange formalism resulting in:

$$f_{PRBM} = \frac{1}{2\,\pi}\sqrt{\frac{c_t}{J_{mp} + m_{mp}\, l_{mp}^2}} \tag{1}$$

The length l_{mp} represents the distance between the axis of rotation and the center of gravity of the moving part. The properties, e.g., mass and inertia are defined in Sect. 2.1. To determine the stiffness of the torsional spring c_t, we used Eq. (2) from [12]:

$$c_t = k_{M1}\, E\, \beta_b\, \beta_l^{(-k_{M2})}\, \beta_h^{(2+k_{M2})}\, H^3 \tag{2}$$

The coefficients k_{M1}, k_{M2}, β_b, β_l, and β_h were selected following [12] and are summarized in Table 4 for the three single flexure hinge specimens, and also, for the parallel-crank mechanism.

Table 4. Parameters of the flexure hinge specimen for a semi-circular contour acco. to [12].

Parameter	k_{M1}	k_{M2}	β_b	β_h	β_l
Value	107.9e-3	0.52	$\frac{w}{H}$	$\frac{h}{H}$	$\frac{2R}{H}$

Finite Element Models
We used two different approaches for the finite element analysis of the specimens. The first model is a 3D finite element model, generated and solved by the software ANSYS Workbench 2020 R1, abbreviated as ANSYS in the following. The need for accurate and computational efficient models for optimization or control applications lead to approaches for model order reduction [14, 15].

The authors of [14] introduced finite beam elements (FBE) of variable cross section for this purpose, which were used in our second approach. The rigid sections of the specimens (moving part in Fig. 2 and 1, 2, 3 in Fig. 3) were modeled with six 2-node beam elements of constant cross section, neglecting minor recesses and bores. Each notched sections of the models consists of six 3-node beam elements of variable cross section according to the circular contour.

3 Results and Discussion

The flexure hinge specimens and a compliant parallel-crank mechanism were investigated based on the proposed free oscillation measurement.

The first eigenfrequency is determined using a chromatic-confocal distance sensor and the extraction of the frequency spectrum by fast Fourier transform (FFT) of the measured oscillation, see Fig. 4. Due to the properties of the flexure hinge, the amplitude spectrum shows a sharply defined peak at the first eigenfrequency, which is the bending mode of the flexure hinge about the z-axis.

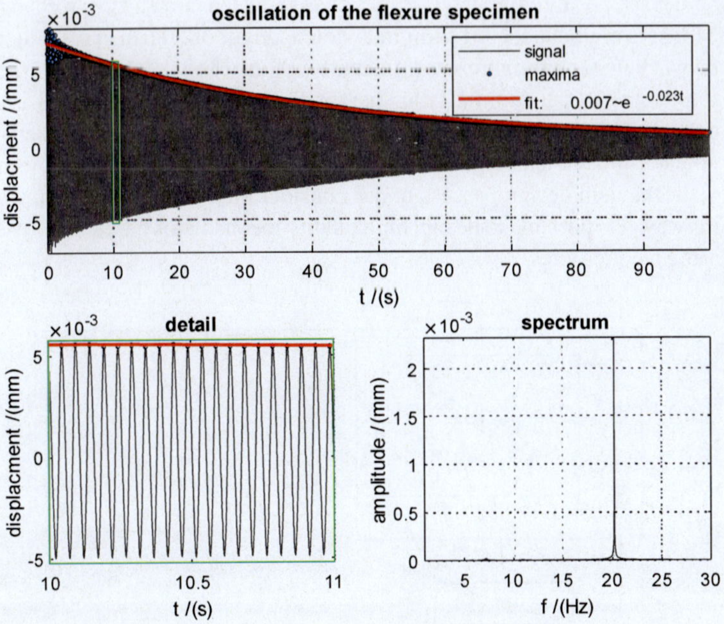

Fig. 4. Oscillation measurement and extraction of frequency spectrum by FFT for the flexure hinge with 50 μm minimal notch height.

The experimental determination of the first eigenfrequency is influenced by energy dissipation to a certain extend. In the present case, energy dissipation at the oscillating flexure hinge occurs by viscous damping of the surrounding air and internal friction within the flexure hinge material. According to theory, the damped eigenfrequency is lower than the undamped one. To preserve consistency with the models, the measured eigenfrequency needs to be corrected to an undamped measured eigenfrequency. This can be accomplished by determining of the decay coefficient δ of the measured oscillation in Fig. 4 with the envelope $Ae^{-\delta t}$. The decay coefficient of the exemplary oscillation in Fig. 4 was determined to $\delta = 0.0231\ \mathrm{s}^{-1}$ which results in a weakly damped system with the damping ratio of $\vartheta \approx 1.1 \times 10^{-3}$. The influence of this weak damping on the eigenfrequency is much smaller than the frequency resolution of the measurement, thus, it was neglected.

The measured eigenfrequencies and the results of the modal analysis of the flexure hinge specimens are presented in Fig. 5 and Table 5. To a large extend, the deviation between the presented models can be explained by the assumption of plane stress in the FBE and the PRBM that is not appropriate for the present flexure hinge geometry [13]. The measurement results are only slightly larger than the ANSYS results, which agrees with dimensional measurements of the minimal notch height. Dimensionally all three hinges are exceeding the nominal value by approximately 1 μm. This agreement is remarkable, since the eigenfrequency of the flexure hinge comprises several physical parameters of the flexure hinge. Among those are the dimensional parameters, especially the minimal notch height, and the material parameters Young's modulus, Poisson number and density. A minimal notch height change of 1 μm results in a change of the rotational stiffness of about 5% for a minimal notch height of 50 μm. This indicates that also surface waviness and roughness significantly influence the rotational stiffness. Due to the great influence of the minimal notch height, this parameter can be used as a substitute parameter representing manufacturing deviations. The difference between nominal and substitute minimal notch height is then directly usable as machining allowance for a subsequent mechanism design process under consideration of the manufacturing capabilities. This way, it is possible to design high quality mechanisms based on cost-efficient manufacturing technologies.

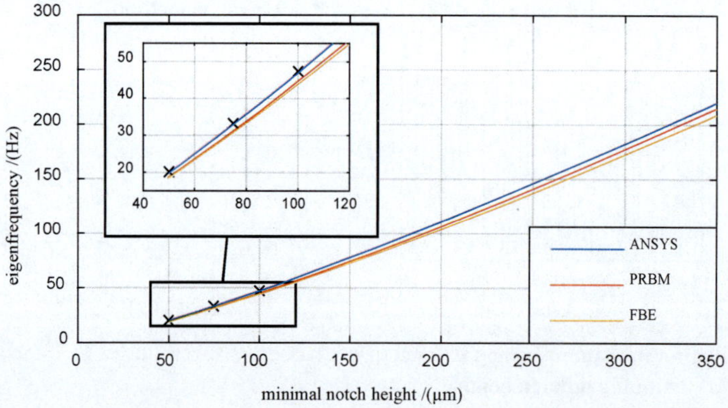

Fig. 5. Comparison between measured (crosses) and predicted eigenfrequencies of the flexure hinge specimen. The curves are power functions over h with slightly different exponents: ANSYS $\approx$ 1.24; PRBM $\approx$ 1.26; FBE $\approx$ 1.25.

The parallel crank compliant mechanism is not directly comparable to the single flexure specimen since the flexure hinge geometry is different and the manufacturing was carried out by another wire-EDM manufacturer. Nevertheless, the results showed good agreement between model and experiment. Accordingly, the presented measurement principle can be applied to compliant mechanisms with arbitrary flexure hinge shapes to determine the *as manufactured* condition. The lower measured eigenfrequency of the mechanism compared to the model with ideal geometry indicates that a certain amount of material is missing in the relevant zone of the flexure hinge. This can either be explained by deviations of the flexure hinge thickness, overlain waviness, and roughness or changes

Table 5. Measured and predicted first eigenfrequencies of specimens.

Specimen	PRBM	FBE	ANSYS	Measurement	h_{subs} (ANSYS)
Flexure hinge ($h = 50\,\mu m$)	18.47 Hz	18.28 Hz	19.74 Hz	20.19 Hz	51.49 μm
Flexure hinge ($h = 75\,\mu m$)	30.78 Hz	30.28 Hz	32.75 Hz	33.25 Hz	75.95 μm
Flexure hinge ($h = 100\,\mu m$)	44.23 Hz	43.43 Hz	46.90 Hz	47.47 Hz	100.83 μm
Parallel crank ($h = 300\,\mu m$)	20.05 Hz	20.03 Hz	20.70 Hz	19.58 Hz	286.77 μm

to the surface material properties. The determination of a mean substitute minimal notch height and its deviation to the nominal value enables an adaption of the mechanical models and a better agreement between model and experiment under consideration of the realized surface integrity. In case of the parallel crank mechanism, a machining allowance of about +13 μm would yield the nominal eigenfrequency of the mechanism.

4 Conclusions and Outlook

This paper raises concern about manufacturing influences on the design of precision mechanisms and promotes its early consideration in the design phase of flexure hinge-based mechanisms. A simple and easy to implement method for the a priori determination of deviations to the projected mechanical properties of flexure hinges is presented. Single flexure hinge specimens are manufactured using the same method/machine in advance of the specific manufacturing process of an entire mechanism. The properties of the specimen and their deviation to the nominal values are determined by measuring the free oscillation of single flexure hinges. Either the mechanism design is adapted to the determined values using e.g. additional machining allowance or the manufacturing process is optimized until the nominal values are achieved.

Following the method and presuming a stable manufacturing process, all practical influences on a specific flexure hinge manufacturing process and the resulting surface integrity are considered in the design phase of the compliant mechanism itself. In case the mechanism design is based on the same model that is used for the deviation determination of the single flexure hinge specimen, the model itself and the assumed material parameters will not significantly influence the results. Hence, all investigated models are considered suitable for the presented characterization approach This provides the opportunity for an efficient mechanism design process with smaller deviations to the projected mechanical properties. After all, it accelerates the mechanism development process based on simple flexure hinge specimens. Another application is the analysis of yet manufactured mechanisms whose *as manufactured* properties can be determined to enable a refined controller design. The proposed method for the design process will be applied to an exemplary design project to determine its practical benefit.

Acknowledgements. The authors gratefully acknowledge the support of the German Research Foundation (DFG) under grant numbers TH 845/5-2 and TH 845/7-2.

References

1. Howell, L.L., Magleby, S.P., Olsen, B.M. (eds.): Handbook of Compliant Mechanisms. Wiley, Chichester (2013)
2. Cosandier, F., Henein, S., Richard, M., Rubbert, L. (eds.): The Art of Flexure Mechanism Design. EFPL Press; CRC Press Taylor & Francis Group, Lausanne, Boca Raton (2017)
3. Lobontiu, N.: Compliant Mechanisms: Design of Flexure Hinges. CRC Press, Boca Raton (2003)
4. Paros, J.M., Weisbord, L.: How to design flexure hinges. Mach. Des. **25**(11), 151–156 (1965)
5. Darnieder, M., Pabst, M., Wenig, R., Zentner, L., Theska, R., Fröhlich, T.: Static behavior of weighing cells. J. Sensors Sensor Syst. **7**(2), 587–600 (2018)
6. Davim, P.J. (ed.): Surface integrity in machining. Springer, London (2010)
7. Lobontiu, N.: Lumped-parameter inertia model for flexure hinges. Mech. Des. Struct. Mach. **32**(1), 73–100 (2004)
8. Dirksen, F., Lammering, R.: On mechanical properties of planar flexure hinges of compliant mechanisms. Mech. Sci. **2**, 109–117 (2011)
9. Ling, M., Cao, J., Jiang, Z., Lin, J.: A semi-analytical modeling method for the static and dynamic analysis of complex compliant mechanism. Precision Eng. **52**, 64–72 (2018)
10. Ling, M., Howell, L.L., Cao, J., Chen, G.: Kinetostatic and dynamic modeling of flexure-based compliant mechanisms: a survey. Appl. Mech. Rev. **72**(3), 030802 (2020)
11. Linß, S., Schorr, P., Zentner, L.: General design equations for the rotational stiffness, maximal angular deflection and rotational precision of various notch flexure hinges. Mech. Sci. **8**(1), 29–49 (2017)
12. Torres Melgarejo, M.A., Darnieder, M., Linß, S., Zentner, L., Fröhlich, T., Theska, R.: On modeling the bending stiffness of thin semi-circular flexure hinges for precision applications. Actuators **7**(4), 86 (2018)
13. Linß, S., Gräser, P., Henning, S., Harfensteller, F., Theska, R., Zentner, L.: Synthesis method for complaint mechanisms of high-precision and large-stroke by use of individually shaped power function flexure hinges. Advances in Mechanism and Machine Science. Mechanisms and Machine Science 73, pp. 1569–1579. Springer, Cham (2019)
14. Friedrich, R., Lammering, R., Rösner, M.: On the modeling of flexure hinge mechanisms with finite beam elements of variable cross section. Precision Eng. **38**, 915–920 (2014)
15. Rösner, M., Lammering, R., Friedrich, R.: Dynamic modeling and model order reduction of compliant mechanisms. Precision Eng. **42**(1), 85–92 (2015)

Optimization of Compliant Path-Generating Mechanisms Based on Non-linear Analytical Modeling

S. Henning[1](✉), S. Linß[1], P. Gräser[2], J. D. Schneider[2], R. Theska[2], and L. Zentner[1]

[1] Compliant Systems Group, Technische Universität Ilmenau, 98693 Ilmenau, Germany
stefan.henning@tu-ilmenau.de

[2] Precision Engineering Group, Technische Universität Ilmenau, 98693 Ilmenau, Germany

Abstract. Monolithic compliant mechanisms are often used in precision engineering applications for path-generating tasks due to their many advantages. They are mostly realized with concentrated compliance in form of notch flexure hinges and achieve their motion due to bending of the hinges. This contribution presents the non-linear analytical modeling of compliant mechanisms with power function-based notch flexure hinges and their efficient optimization of the elasto-kinematic path-generating properties using MATLAB. Different planar mechanisms are analytically characterized with the theory for large deflections of curved rod-like structures. A verification of the analytical model is exemplified by FEM simulations for a four-hinge WATT mechanism as a point guidance mechanism and for a 12-hinge pantograph mechanism as a plane guidance mechanism. Further, the exponents of the power function contours for each hinge are individually optimized on the example of an EVANS and a ROBERTS mechanism. This is achieved with the goal of minimizing the straight-line deviation of their coupler points realizing a stroke of 10 mm.

Keywords: Compliant mechanism · Flexure hinge · Analytical model · Large deflection · FEM

1 Introduction

Compliant mechanisms are increasingly used in precision engineering, measuring and microtechnology [1]. Due to their advantages such as frictionless and clearance-free motion, they are ideally suited for guiding and precision tasks [2]. In those applications, they mostly comprise concentrated compliances in the form of notch flexure hinges [3]. The motion behavior of a compliant mechanism can mainly be influenced by the geometry of the flexure hinge contour. Various notch contours exist in literature. In the scope of this paper, the highly adaptable power function-based contours are investigated [4]. Due to the monolithic manufacturing, there is no defined axis of rotation of the hinges which results in deviations of the guidance of specific points or planes compared to the

L. Zentner and S. Strehle (Eds.): MAMM 2020, MMS 96, pp. 25–35, 2021.
https://doi.org/10.1007/978-3-030-61652-6_3

ideal rigid-body model [5]. The authors aim with this paper is to counteract this fact by optimizing the flexure hinge contours for the reduction of guidance deviations of chosen mechanisms. However, the analysis and synthesis of compliant mechanism is a challenging task due to non-linearities caused by large deflections. Mostly time-consuming FEM-based approaches are applied in an iterative analysis process [6]. Therefore, an analytical model is presented in this contribution which allows a fast and accurate evaluation of important elasto-kinematic properties of complex compliant mechanisms and which is suitable for optimization procedures.

2 Analytical Model

In this section, a non-linear analytical model is derived and applied to continuous and branched compliant mechanisms based on a section wise implementation.

2.1 Large Deflections of Rod-Like Structures

For analytically modeling the investigated compliant mechanisms, a non-linear approach based on the theory for large deflections of rod-like structures is considered. Therein, the equilibrium equations are formulated in the deflected state of a continuous rod [7]. Several assumptions are made like the validity of HOOKE's law, St. VENANT's principle and BERNOULLI's hypotheses. On the example of a single rod with one fixed end and another free end (Fig. 1), a system of differential equations is derived, where only bending is considered.

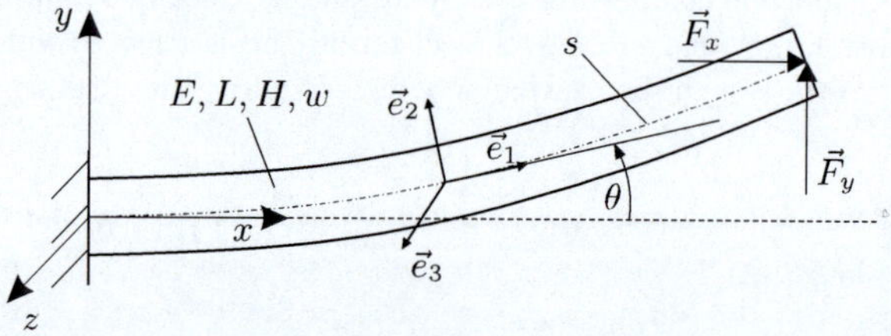

Fig. 1. A rod in its deflected state, loaded by the external direction-constant forces $\vec{F}_x$ and $\vec{F}_y$.

The rod with the length L, the height H and the width w is loaded by two external forces $\vec{F}_x$ and $\vec{F}_y$ and deflected with the angle θ around the $\vec{e}_3$ axis. To describe the deflected state of the rod, the following differential equations are obtained along the rod axis s:

$$\frac{\mathrm{d}Q_x}{\mathrm{d}s} = 0, \tag{1}$$

$$\frac{\mathrm{d}Q_y}{\mathrm{d}s} = 0, \tag{2}$$

$$\frac{\mathrm{d}M_z}{\mathrm{d}s} = Q_x \sin\theta - Q_y \cos\theta, \tag{3}$$

$$\frac{\mathrm{d}\theta}{\mathrm{d}s} = \kappa, \text{ with } \kappa = \frac{M_z}{EI_z} + \kappa_0, \tag{4}$$

$$\frac{\mathrm{d}x}{\mathrm{d}s} = \cos\theta, \tag{5}$$

$$\frac{\mathrm{d}y}{\mathrm{d}s} = \sin\theta. \tag{6}$$

Considering the fixed and free ends of the rod in Fig. 1, the following boundary conditions result:

$$Q_x(L) = F_x, Q_y(L) = F_y, M_z(L) = 0, \theta(0) = 0, x(0) = 0, y(0) = 0. \tag{7}$$

The system of differential equations represents a boundary value problem. Regarding non-constant cross-sections as in the case of compliant mechanisms with notch flexure hinges, the system of differential equations is heteronomous, and no closed-form analytical solution can be carried out. Therefore, numerical methods are necessary to solve the given problem. A conventional solution for those problems is the shooting method, in which a guess is made for all unknown parameters and then varied until the boundary conditions are matched within a specific tolerance (e.g. [8]).

2.2 Modeling of Compliant Mechanisms as Continua

For applying the given approach, it is required that the investigated compliant mechanisms are modeled as continuous rods. Along the path of the rod-axis s of a compliant mechanism, changes in the cross-section or curvature may occur (Fig. 2a). A compliant mechanism may, therefore, be divided into several sections as demonstrated in Fig. 2a. In summary, any continuous compliant mechanism can be built from three different building blocks – beam segments of constant cross-section (Fig. 2b), curved beam segments (Fig. 2c) or flexure hinges with their specific notch contour (Fig. 2d). Curved beam segments are implemented into the curvature in Eq. (4) and defined by a curve radius R according to Eq. (8) depending on their orientation.

$$\kappa_0 = \pm\frac{1}{R} \tag{8}$$

The flexure hinge contour is implemented into the second moment of area I_z by specifying the contour height function h_c along the length of the hinge from $s = 0$ to $s = L$:

$$h_c(s) = h + 2^n \cdot \frac{H - h}{L^n} \cdot \left| s - \frac{L}{2} \right|^n. \tag{9}$$

However, compliant path-generating mechanisms often contain an additional branch at the coupler or other points where links or hinges are connected. In those cases, the above theory is not valid because of the resulting discontinuities. Alternatively, the mechanism can be divided into several continuous rods first and then be further divided into the three different building blocks of Fig. 2c. Afterwards, transition conditions need to be formulated to connect multiple continuous rods with one another. Using the shooting method, the theory can be applied for these mechanisms, too [9].

Regarding notch flexure hinges, the investigations of this paper are focused on power function-based contours (Fig. 2d). These provide a good compromise between precision and strain distribution by adjusting the exponent n [4].

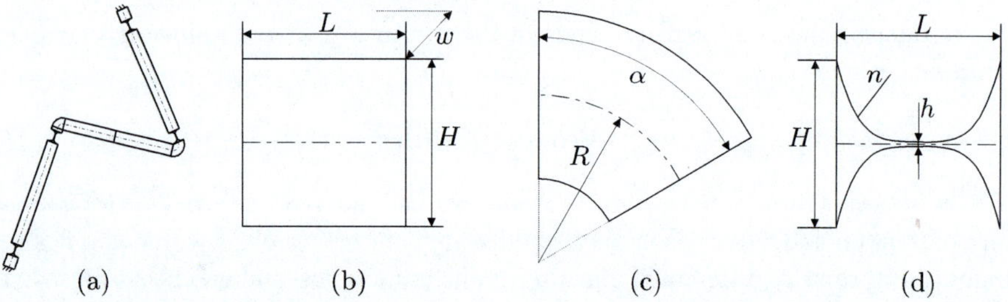

Fig. 2. Segments of a compliant mechanism: (a) compliant WATT mechanism divided into 13 segments, (b) beam segment with constant cross-section, (c) curved beam segment, (d) power function-based flexure hinge contour segment.

3 Verification of the Analytical Model by FEM Simulation

To verify the results obtained with the proposed analytical model, two example path-generating mechanisms are investigated by means of the implemented analytical approach and by non-linear 3D-FEM simulations. Both are modeled to be made from the typically used aluminum alloy EN AW 7075 with a YOUNG's modulus of 72 GPa and an admissible elastic strain of $\varepsilon_{\text{adm}} = 0.5\%$.

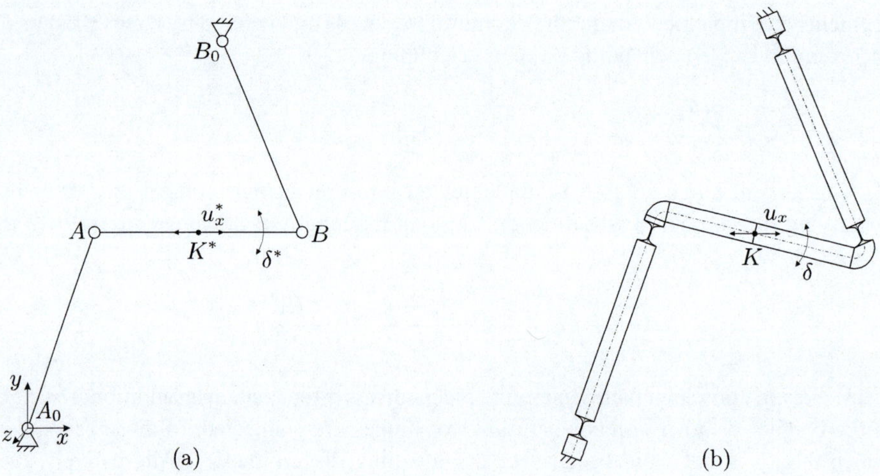

Fig. 3. Point guidance WATT mechanism: (a) rigid-body model, (b) compliant mechanism.

First, the compliant WATT mechanism realizes the guidance of a coupler point K on an approximated straight horizontal line while a rotation of the coupler occurs (Fig. 3a). This mechanism does not contain any branches and is continuous, consisting of beam segments with constant cross-section, curved beam segments and power function-based flexure hinge contours.

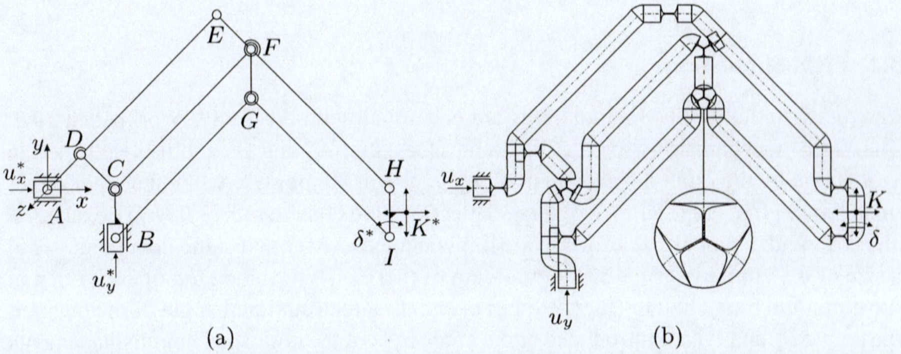

Fig. 4. Plane guidance pantograph mechanism: (a) rigid-body model, (b) compliant mechanism.

Second, a more complex compliant pantograph-based mechanism after [10] is considered to represent the guidance of a plane in x and y-direction while in the ideal rigid-body model no rotation occurs (Fig. 4a). The pantograph mechanism consists of eleven links, twelve hinges, implies two input prismatic joints in x- and y-direction and magnifies the input displacement to the output plane. By increasing the stroke to the output, the properties of highly reproducible actuators, which usually only provide a small range of displacement, can be extended to a larger stroke. Clearly, six branching points are present within the compliant version of this mechanism so that transition conditions need to be considered. Furthermore, three of those branching points are novel planar two-axis flexure hinges with coincident rotation axes. Here, they are designed using half models of the power function-based contours in Fig. 2d, for easy implementation with an exponent of $n = 14$ according to the hinge angles and are interconnected at the branching point.

3.1 Analytical Solution

The compliant Watt mechanism is deflected by an input displacement of $u_x = 10\,\text{mm}$ while the compliant pantograph mechanism is deflected by two displacements of $u_x = u_y = 2.5\,\text{mm}$ while the motion transmission ratio is around 5. The analytical solution of both mechanisms is performed within MATLAB® by numerically solving the system of differential equations in Eq. (1)–(7). A combination of a four-step RUNGE-KUTTA method (*ode45*) with a shooting method (*fsolve*) is performed until all boundary and transition conditions are matched. Once a solution is obtained, it is possible to evaluate the displacement of the point K using the results of Eq. (5) and (6). The rotational angle δ of point K is given by the solution of Eq. (4) and the reaction forces can be obtained with

Eq. (1) and (2). The results of the equations can also be used for further processing of important elasto-kinematic parameters. For example, the maximum elastic strain $\varepsilon_{\max}$ is calculated by Eq. (10) with η being the distance from the rod axis to the outer fiber.

$$\varepsilon = \left| \frac{M_z}{EI_z} \cdot \eta \right| \tag{10}$$

3.2 FEM Simulation

The quasi-static 3D FEM simulations are performed with ANSYS® Workbench 19.2. The WATT mechanism is fixed at both end faces after A_0 and B_0 and it is deflected in x-direction by the input displacement of $u_{xK} = 10$ mm in point K while the other directions remain free (Fig. 5a). Also, large deflections are considered. *Solid186* hexahedral elements with a global size of 1 mm are implemented and refined at the flexure hinges to a size of 0.15 mm to ensure that a minimum of two elements are on top of each other at the minimum hinge height (Fig. 5b). For calculating the rotational angle of the coupler, the points P_1 and P_2 are introduced at the outer ends of the coupler. Their displacements are evaluated and used to calculate the rotation angle.

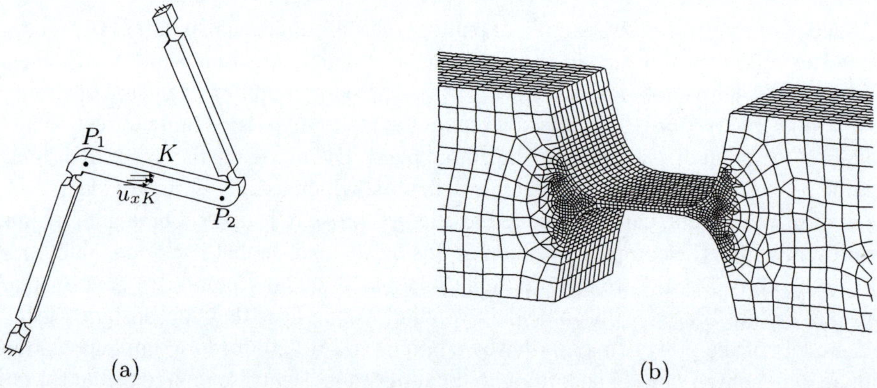

Fig. 5. 3D FEM model on the example of the compliant WATT mechanism with sixth order power function-based flexure hinge contours: (a) boundary conditions and important points, (b) detailed mesh of one flexure hinge.

The pantograph mechanism is modeled analogously and deflected at the x-slider by the displacement $u_x = 2.5$ mm while the y- and z-directions remain inhibited. The same is done at the y-slider by the displacement $u_y = 2.5$ mm while the x- and z-directions remain inhibited.

3.3 Results and Discussion

Following, the results of both example mechanisms are outlined for the given input deflections. According to the relative hinge rotation angles, the investigations are done with the individually shaped flexure hinge exponent n [4] given in Table 1 and 2. The power function exponents are evaluated using design graphs according to the maximum occurring rotation angles φ of the ideal rigid-body model for the given displacements.

Table 1. Kinematics and chosen exponents n of the WATT mechanism.

Parameter	A_0	B_0	A	B	K
x	0	98.2	34.2	138.2	86.2
y	0	185.6	94	94	94
φ	6.3	6.3	11.5	1.2	–
n	5	5	11	2	–

Table 2. Kinematics and exponents n of the pantograph mechanism.

Parameter	A	B	C	D	E	F	G	H	I	K
x in mm	0	40	40	20	100	120	120	200	200	210
y in mm	0	−28	0	−20	100	80	52	0	−28	−14
φ in °	6.5	0.4	13.6	6.8	6.8	13.6	13.6	0.4	0.4	–
n	5	5	14	6	6	14	14	5	5	–

Table 3. Analytical and FEM simulation results of the WATT and pantograph mechanisms.

Method	u_{xK} in mm	u_{yK} in mm	δ in °	F_x in N	F_y in N	ε_{max} in %
WATT mechanism deflected by $u_{xK} = 10$ mm						
FEM	10	0.2911	5.11	1.38	–	0.54
Matlab	10	0.2914	5.11	1.52	–	0.49
Deviation	0%	0.01%	0%	10.14%	–	−9.26%
Pantograph mechanism deflected by $u_{xA} = u_{yB} = 2.5$ mm						
FEM	−10.4730	11.6010	1.09	7.53	2.04	0.56
Matlab	−10.4638	11.6193	1.07	7.08	1.93	0.51
Deviation	−0.09%	0.16%	−1.83%	−5.98%	−5.39%	−8.93%

The results for the x- and y-displacements, the rotation angle of the coupler, the reaction forces and the overall maximum elastic strain of both mechanisms are given in Table 3. Furthermore, the relative deviations of the analytical results from the 3D FEM simulation results are outlined in the table, too.

In summary, the FEM and analytical results are in good correlation. Very small deviations occur regarding the displacement of arbitrary points. Even angular rotations are well matched between both methods. The highest deviations are obtained in terms of forces and consequently for the maximum elastic strain. Since, in case of the analytical model, only bending is considered, the differences can further be reduced when other effects like shear or lateral contraction are considered [11].

4 Analytical Optimization of Example Compliant Path-Generating Mechanisms

Due to the verification of the analytical results by FEM simulations for a continuous and a branched compliant mechanism, the following optimization is performed with the proposed analytical model. The optimization is carried out in the form of comprehensive parametric studies. The main advantages of the analytical model over the FEM simulation is the fast modeling of the geometry and the fast analysis within a few seconds. Therefore, the model is also suitable for synthesis.

Following, two-point guidance mechanisms, with straight-line deviations in the micrometer range regarding the ideal rigid-body model of the mechanisms, are examined. An optimization of the flexure hinge contours is carried out for an EVANS (Fig. 6a) and a ROBERTS mechanism (Fig. 6b). Both mechanisms represent four-bar linkages with a branched coupler. The EVANS mechanism is deflected by $u_{xK} = -10\,\text{mm}$ and the ROBERTS mechanism by $u_{xK} = 10\,\text{mm}$ based on previous investigations [12]. The ideal rigid-body models result in straight-line deviations of $d_s = -55.66\,\mu\text{m}$ (EVANS mechanism) and $d_s = -24.99\,\mu\text{m}$ (ROBERTS mechanism).

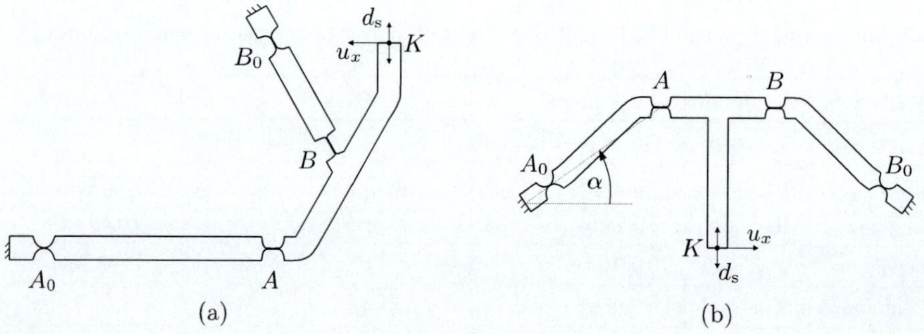

Fig. 6. Compliant models of the investigated point guidance mechanisms: (a) EVANS mechanism ($\overline{A_0A} = \overline{AK} = 100\,\text{mm}$, $\overline{AB} = \overline{BK} = 50\,\text{mm}$, $\overline{A_0B_0} = 132,3\,\text{mm}$), (b) ROBERTS mechanism ($\overline{A_0A} = \overline{B_0B} = 66.6\,\text{mm}$, $\overline{AB} = 56.6\,\text{mm}$, $\overline{A_0B_0} = 165.7\,\text{mm}$, $\overline{AK} = \overline{BK} = 73.6\,\text{mm}$, $\alpha = 35°$).

The goals of the subsequent optimizations are twofold. On one hand, the minimization of the straight-line deviation (d_s) is of great interest. On the other hand, the maximum elastic strain must be lower than the admissible elastic strain of $\varepsilon_{\text{adm}} = 0.5\%$. This will be investigated by changing the exponent of the power function-based flexure hinge contours in the form of a parametric study. The objective of the optimization will be the minimization of the straight-line deviation d_s. A constraint for the valid results is chosen to be the maximum elastic strain as $\varepsilon_{\max} \leq \varepsilon_{\text{adm}}$.

In a first investigation, the minimum exponents using natural numbers, that will represent the lower limits of the parametric study, are evaluated for each flexure hinge which largely depends on the rotational angle. Second, the exponents for each hinge are varied, also with natural numbers. The upper limit of the exponent is chosen to be $n = 20$. For each combination of exponents, the straight-line deviation is evaluated until an optimum satisfying the given constraint is found. Afterwards, a finer search is done near the obtained optimum with the use of rational exponent numbers in steps of 0.1 until a new optimum can be found. The results of the optimizations based on the objective and constraint are outlined in Table 4. Lower exponents of the power function result in a more concentrated compliance and generally in more precise flexure hinges when isolated. However, they do not indicate the optimum regarding the straight-line deviation of the mechanisms. Interestingly, it was not possible to significantly reduce the straight-line deviation of the EVANS mechanism by not exceeding the admissible elastic strain. However, if the elastic strain was not an issue it would be possible to get much lower. In case of the ROBERTS mechanism, the straight-line deviation could be drastically reduced to almost zero which is significantly lower than the deviation of the rigid-body model.

Table 4. Optimal power function exponents for the EVANS and the ROBERTS mechanism.

	n_{A0}	n_A	n_B	n_{B0}	F_{xK} in N	d_s in µm	ε_{max} in %
EVANS mechanism deflected by $u_{xK} = -10$ mm							
Minimum exponents	3	7	16	5	−1.55	−93.4589	0.50
Optimum for natural numbers	10	14	16	5	−1.34	−91.4449	0.50
Optimum for rational numbers	10.5	14.5	16.1	4.6	−1.36	−91.2225	0.50
ROBERTS mechanism deflected by $u_{xK} = 10$ mm							
Minimum exponents	3	9	8	3	1.37	−1.4816	0.48
Optimum for natural numbers	8	9	8	7	1.18	0.0003	0.47
Optimum for rational numbers	7.7	8.6	8.1	6.9	1.19	0.0001	0.48

In general, the results of the EVANS and ROBERTS mechanism from previous studies based on 3D FEM simulations [12] could be reproduced with the analytical model of this research. Concerning the ROBERTS mechanism, it has been possible to achieve a further improvement of the straight-line deviation by simultaneously considering the elastic strain constraint with rational numbers as exponents for the power function. Also, the straight-line deviation of the EVANS mechanism has been further reduced even when considering the elastic strain.

5 Conclusions

The analytical investigations of this paper are based on the theory for large deflections of rod-like structures. It has been shown that the obtained results could be verified by 3D FEM simulations for a four-hinge WATT mechanism and a 12-hinge pantograph mechanism. The approach can easily be applied and is timesaving for comprehensive parametric studies and optimizations and thus also well suited for the synthesis of compliant mechanisms. Furthermore, it was possible to analytically model planar two-axis flexure hinges with coincident rotation axes as branched hinges. Moreover, an optimization in the form of parametric studies has been performed on the example of an EVANS and a ROBERTS mechanism to minimize the straight-line deviation by changing the exponents of the flexure hinges, considering the elastic strain. With the application of rational numbers, the guidance errors could be further reduced. A further optimization of compliant guidance mechanisms independent of the rigid-body model is possible, whereby even a complete compensation of the straight-line deviation for a certain output displacement can be achieved. In future research, the analytical investigations will be carried out for spatial compliant mechanisms, too.

Acknowledgements. The authors would like to gratefully acknowledge the support of the German Research Foundation (DFG) under grant number ZE 714/10-2.

References

1. Howell, L.L., Magleby, S.P., Olsen, B.M. (eds.): Handbook of Compliant Mechanisms. Wiley, Chichester (2013)
2. Cosandier, F., Henein, S., Richard, M., Rubbert, L. (eds.): The Art of Flexure Mechanism Design. EFPL Press; CRC Press Taylor & Francis Group, Lausanne, Boca Raton, Fl (2017)
3. Lobontiu, N.: Compliant Mechanisms: Design of Flexure Hinges. CRC Press, Boca Raton (2003)
4. Linß, S., Gräser, P., Henning, S., Harfensteller, F., Theska, R., Zentner, L.: Synthesis method for compliant mechanisms of high-precision and large-stroke by use of individually shaped power function flexure hinges. In: Uhl, T. (ed.) Advances in Mechanism and Machine Science. IFToMM WC 2019. Mechanisms and Machine Science, vol. 73, pp. 1569–1578. Springer, Cham (2019)
5. Venanzi, S., Giesen, P., Parenti-Castelli, V.: A novel technique for position analysis of planar compliant mechanisms. Mech. Mach. Theory **40**(11), 1224–1239 (2005)
6. Friedrich, R., Lammering, R., Heurich, T.: Nonlinear modeling of compliant mechanisms incorporating circular flexure hinges with finite beam elements. Precis. Eng. **42**, 73–79 (2015)
7. Zentner, L., Linß, S.: Compliant Systems: Mechanics of Elastically Deformable Mechanisms, Actuators and Sensors. De Gruyter, Berlin (2019)
8. Ascher, U.M., Petzold, L.R.: Computer Methods for Ordinary Differential Equations and Differential-Algebraic Equations. Society for Industrial and Applied Mathematics, Philadelphia (1998)
9. Henning, S., Linß, S., Vollrath, T., Zentner, L.: Elasto-kinematic modeling of planar flexure hinge-based compliant mechanisms incorporating branched links. In: Uhl, T. (ed.) Advances in Mechanism and Machine Science. IFToMM WC 2019. Mechanisms and Machine Science, vol. 73, pp. 1599–1608. Springer, Cham (2019)

10. Horie, M., Hoshikawa, Y., Kamiya, D.: Optimum design of mass distribution of the injection molding pantograph mechanism with constant output link orientation. In: Ananthasuresh, G.K., Corves, B., Petuya, V. (eds.) Micromechanics and Microactuators 1st MAMM, Mechanisms and Machine Science, vol. 2, pp. 31–39. Springer, Dordrecht (2011)
11. Henning, S., Zentner, L.: Nichtlinear-analytische Modellbildung nachgiebiger Mechanismen unter Berücksichtigung des Querkraftschubs. In: Gössner, S. (ed.) Tagungsband 13. Kolloquium Getriebetechnik, pp. 31–40, Logos, Berlin (2019)
12. Gräser, P., Linß, S., Zentner, L., Theska, R.: Optimization of compliant mechanisms by use of different polynomial flexure hinge contours. In: Kecskeméthy, A., Geu Flores, F., Carrera, E., Elias, D.A. (eds.) Interdisciplinary Applications of Kinematics. Proceedings of the Third International Conference (IAK). Mechanisms and Machine Science, vol. 71, pp. 265–274. Springer, Cham (2019)

Modelling and Investigation of a Compliant Cable-Driven Finger-Like Mechanism

R. Uhlig(✉), L. Zentner, and M. Wolfenstetter

Department of Mechanical Engineering, Compliant Systems Group, Technical University of Ilmenau, Ilmenau, Germany
rene.uhlig@tu-ilmenau.de

Abstract. This paper deals with the modelling of a compliant finger-like mechanism that is actuated by a single cable. The mechanism has a desired elastic deformation behavior which can be adjusted by the mechanism design. The modelling of the compliant mechanism is based on the nonlinear theory of large deformations of curved rods. The known theory is extended by a novel approach of modelling the interaction between the cable and the mechanism itself. In addition, the influence of the gravity is discussed. The resulting model is verified by experiments.

Keywords: Compliant mechanism · Bending shape · Cable-driven · Large deformations of curved rods · Finger-like · Modelling

1 Introduction

The human hand is the natural example of a universal, adaptable gripping mechanism. In the past, there have been many attempts to transfer the principle of human gripping into technical applications (Catalano et al. 2014; Deimel and Brock 2016; Grebenstein et al. 2012; Xu and Todorov 2016).

The gripping function of the human hand is realized by the interaction of the fingers and the thumb. Each finger consists of three phalanges which are rotatably connected with each other. In order to stretch and bend the fingers, each finger has two tendons which are connected to the outer phalanx on the one end and to the muscles in the forearm on the other end. The finger movement is based on muscle contraction that is transmitted via the tendons. Obviously it is possible to grip different shaped objects. Depending on which phalanx the contact to the object to be gripped is made, there is a slight or greater flexion of the finger.

The development of a compliant finger-like mechanism is based on the natural example of the human finger. The finger-like mechanism in this paper has three rigid segments as well which are linked by compliant segments. The compliant segments represent the joints of the mechanism because of their distributed compliance. A guided cable is used to actuate the mechanism (flexion). The stretching of the mechanism is made passively by the compliance of the joint segments.

L. Zentner and S. Strehle (Eds.): MAMM 2020, MMS 96, pp. 36–47, 2021.
https://doi.org/10.1007/978-3-030-61652-6_4

Cable-driven mechanisms are already known (Megaro et al. 2017; Cheng et al. 2013). The special feature of the mechanism described in this paper is the use of a single cable and compliant joints segments. The use of a cable enables the spatial separation of mechanism and actuator. For this reason, it is possible to miniaturize the mechanism independently of the actuator. There are versatile areas of application for the finger-like mechanism ranging from robotics, gripping and endoscopic applications to application as adaptable instruments in the microsurgery.

2 Basic Concept of Compliant Cable-Driven Mechanism

The finger-like mechanism in this paper consists of five segments of bending stiffness S_0, S_I and S_{II} (see Fig. 1). The segments of the stiffness S_I and S_{II} are the compliant segments of the mechanism and represent the joints of mechanism. The bending stiffness S_0 of the remaining three segments is typically much larger than the bending stiffness of the compliant segments S_I and S_{II}. Therefore, the segments S_0 are regarded as comparatively rigid. A cable, which is attached to the outer end of the mechanism at a vertical distance a, is used to actuate the mechanism. To ensure a constant distance between the cable and the mechanism segments, multiple spacers are used to guide the cable. The (external) actuator, which exerts a pulling force to the cable, is not considered in this paper.

In Fig. 1 the mechanism is shown in the non-actuated state a) and the exemplary actuated states b), c) and d). It should be noted, that bending is only considered in plane. In the non-actuated state, the mechanism is straight (comparable to the human stretched finger). In state b) the bending stiffness S_I of the joint segment is assumed to be equal to S_0. Therefor the bending principally occurs in the joint segment S_{II}, which allows a specific movement of the fingertip. In state c), the movement is focused on the segment

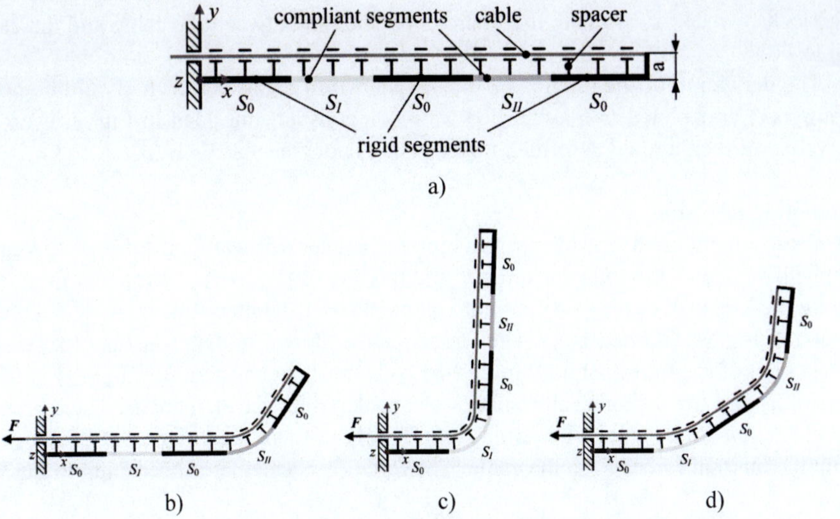

Fig. 1. Basic concept of compliant cable-driven mechanism. a) non-actuated state; actuated states: b) $S_I = S_0 > S_{II}$; c) $S_I < S_0 = S_{II}$; d) $S_I, S_{II} < S_0$

S_I by setting the bending stiffness S_{II} to the value of S_0. Finally, when the bending stiffness of both joint segments S_I and S_{II} is below S_0 – as shown in state d) – bending occurs in both joint segments.

By adjusting the bending stiffness S_I and S_{II} of the joint segments, a wide range of desired bending shapes can be realized. Due to this concept, the mechanism can be adapted to a wide field of applications, as described in Sect. 1. Miniaturization becomes possible due to the separation of the mechanism from the actuator and the possibility of a monolithic production.

3 Model and Simulation

3.1 Mechanical Model

The described mechanism in the previous section is transferred to the mechanical model shown in Fig. 2. The total length of the mechanism in its undeformed configuration is depicted by the parameter l. The five mechanism segments have equal length of $0.2l$. The bending stiffness S_0 of the rigid segments is defined by $S_0 = E_0 \cdot I_{z0}$. The bending stiffness S_I and S_{II} of the joint segments is expressed by the general formula $S_i = c_i \cdot S_0$, whereby c_i is a stiffness factor $0 < c_i \leq 1$ which sets the bending stiffness S_I and S_{II} in reference to the bending stiffness S_0. In the mechanical model the cable is replaced by an equivalent set of mechanical loads. It is assumed that the cable is limp and only transfers tensile forces. Therefore, the cable can be modelled by a follower force F which – together with the lever arm a – creates a moment M at the free end of the mechanism. The spacers, guiding the cable (see Fig. 1a), transfer forces between the cable and the mechanism. Otherwise the cable would lift off during the flexion of the mechanism. In the mechanical model the forces transferred by the spacers are described by a novel approach. Therefor a line load q_2 is introduced which depends on the curvature κ_3 of the mechanism. It is assumed, that there is no friction between the cable and the cable guide.

The derivation of the line load $q_2(\kappa_3)$ is shown in detail below. It should be noted that $q_2(\kappa_3)$ is depicted in a simplified way as a constant line load in Fig. 2, even the curvature of the pictured deformed mechanism is not constant.

Modelling of Cable

As already mentioned, the tensile force of the actuator transmitted through the cable can be modeled as the follower force F and the line load $q_2(\kappa_3)$ as shown in Fig. 2. To derive the line load $q_2(\kappa_3)$ a differential element ds of the mechanism is considered as shown in Fig. 3a (dashed box). The element ds is shown in detail in Fig. 3b. For the description of loads and internal forces the rod coordinate system $\{e_1, e_2, e_3\}$ is used. Regarding Fig. 3b, a single cable spacer is considered for the element ds. The forces on the element ds must be in an equilibrium. Therefore, there is a force F_2, which results from the internal forces F on the cable. The force F_2 is calculated according to Eq. (1).

$$dF_2 = 2F \sin\left(\frac{d\theta_3}{2}\right) = 2F\frac{d\theta_3}{2} = Fd\theta_3 \tag{1}$$

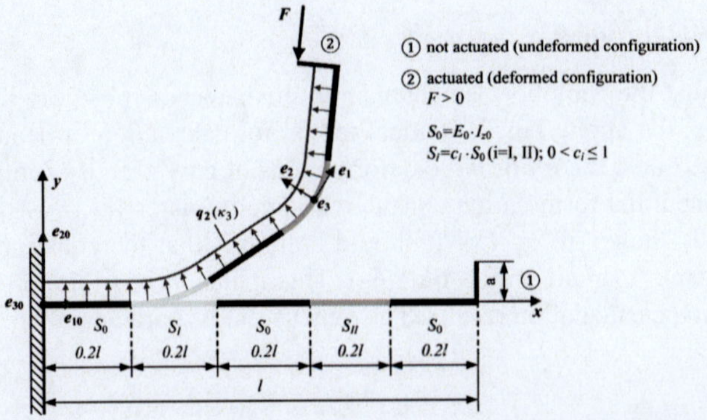

Fig. 2. Model of compliant cable-driven mechanism in actuated and not actuated state.

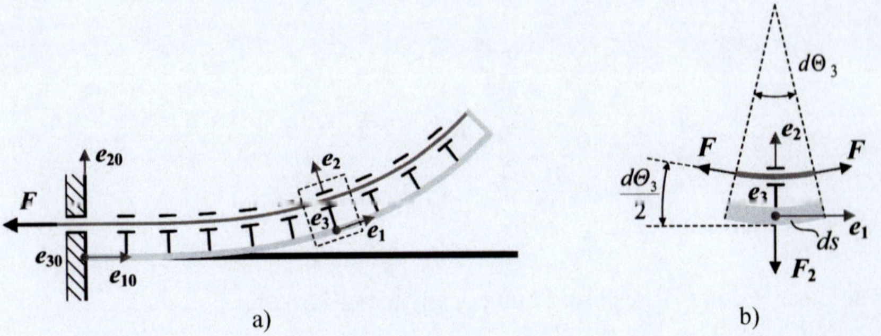

Fig. 3. Modelling of cable: a) overall mechanism; b) differential mechanism element *ds*

Equation (1) is then used to derive Eq. (2) for the calculation of the line load q_2. As can be seen from Eq. (5), the expression $d\theta_3 / ds$ from Eq. (2) describes the curvature κ_3 of the mechanism. When considering the entire mechanism, the cable guides can thus be replaced by the curvature-dependent line load $q_2(\kappa_3)$ as shown in the derived model in Fig. 2.

$$q_2 = \frac{dF_2}{ds} = F\frac{d\theta_3}{ds} = F\kappa_3 \tag{2}$$

Modelling of Gravity

Gravity is a directional force, that is taken into account by the line loads q_{w1} and q_{w2} since the rod coordinate system is considered in the model, see Eq. (5). When gravity acts in the negative y-direction (considering the model in Fig. 2), the components $q_{w1}(e_1)$ and $q_{w2}(e_2)$ are calculated according to Eqs. (3) and (4). Section 4 presents simulation results with and without the influence of gravity. If gravity is neglected, q_w is set to zero.

$$q_{w1} = -\frac{m \cdot g}{l} \sin\theta_3 = -q_w \sin\theta_3 \tag{3}$$

$$q_{w2} = -\frac{m \cdot g}{l} \cos\theta_3 = -q_w \cos\theta_3 \tag{4}$$

3.2 Mathematical Model

It is assumed, that the compliant mechanism in this paper is subject to large deformations. Therefore, the applied mathematical model for calculation of bending shapes is based on the nonlinear theory of large deformations of curved rods (Zentner 2014). As already mentioned, the formulation considers rod coordinates $\{e_1, e_2, e_3\}$. The mathematical model is shown in Eq. (5). It should be noted, that the equations use dimensionless parameters marked with a tilde (~). This makes it possible to derive general statements from the calculation results. The dimensionless parameters can be calculated using Eq. (6).

$$
\begin{aligned}
&\frac{d\tilde{Q}_1}{d\tilde{s}} - \tilde{\kappa}_3\tilde{Q}_2 + \tilde{q}_{w1} = 0 \quad && \frac{d\tilde{Q}_2}{d\tilde{s}} + \tilde{\kappa}_3\tilde{Q}_1 + \tilde{\kappa}_3\tilde{F} + \tilde{q}_{w2} = 0 \\
&\frac{d\tilde{M}_3}{d\tilde{s}} + \tilde{Q}_2 = 0 && \tilde{M}_3 = \tilde{\kappa}_3 \\
&\frac{d\theta_{e3}}{d\tilde{s}} - \tilde{\kappa}_3 = 0 && \\
&\frac{d\tilde{x}}{d\tilde{s}} - \cos\theta_{e3} = 0 && \frac{d\tilde{y}}{d\tilde{s}} - \sin\theta_{e3} = 0
\end{aligned}
\tag{5}
$$

$$
\begin{aligned}
\tilde{Q} &= \frac{Ql^2}{E_0 I_{z0}} \quad & \tilde{F} &= \frac{Fl^2}{E_0 I_{z0}} \quad & \tilde{M}_3 &= \frac{M_3 l}{E_0 I_{z0}} \\
\tilde{x} &= \frac{x}{l} & \tilde{y} &= \frac{y}{l} & \tilde{q} &= \frac{ql^3}{E_0 I_{z0}}
\end{aligned}
\tag{6}
$$

The Eqs. (5) and (6) are solved numerically using *MATLAB*.

4 Calculation Results

The following chapter shows the results derived from the calculation without and under consideration of gravity. In case the weight of the mechanism is not considered, the parameters q_{w1} and q_{w2} in Eq. (5) are set to zero. The length of the lever arm on the mechanism's free end is taken into account with a constant value of $\tilde{a} = 0.05$ in all calculations. The calculations are made by specifying the tensile force $\tilde{F}$ on the cable and the stiffness factors c_I and c_{II} whereby c_I and c_{II} can take values between 0 and 1. For values of the stiffness factor c_i near 1, the joint segment's bending stiffness S_i is approximately the bending stiffness of the rigid segments S_0. To evaluate the results the distance $\Delta\tilde{s}$ is computed which stands in this context for the travel distance of the cable under the influence the force $\tilde{F}$.

4.1 Bending Shapes Without Consideration of Gravity

In Fig. 4 the calculated bending shapes of the compliant mechanism are shown for values of $\Delta\tilde{s}$ between 0 and 0.24. The weight of the mechanism is not considered. The stiffness factors are set to equal values of $c_I = c_{II} = 0.1$ that means the bending stiffness of the joint segments is 1/10 of the bending stiffness of the rigid segments. For illustration reasons the rigid segments of the mechanism are highlighted by black lines. It is obvious that the curvature of the mechanism increases with rising travel distance $\Delta\tilde{s}$ of the cable.

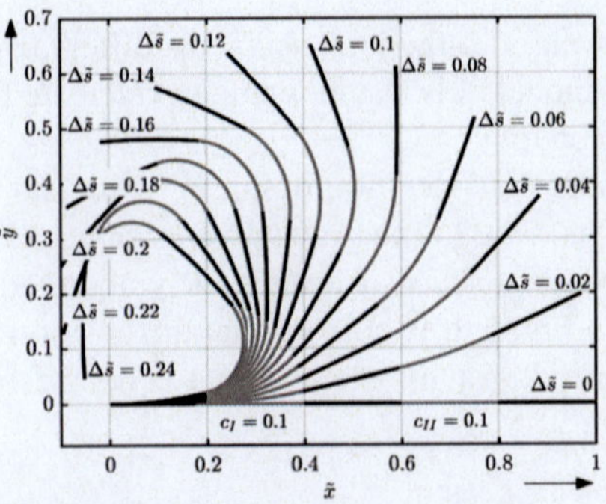

Fig. 4. Calculated bending shapes for $c_I = c_{II}$

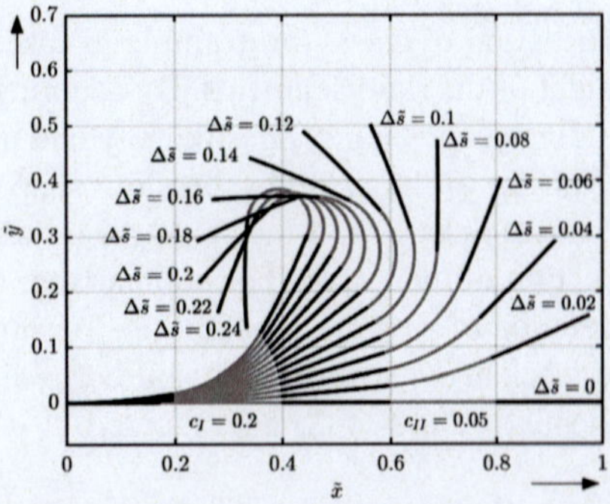

Fig. 5. Calculated bending shapes for $c_I \neq c_{II}$

It should be noted that the corresponding cable force $\tilde{F}$ is proportional to the travel distance of the cable when gravity is not considered.

The resulting bending angle of the joint segments are also proportional to the travel distance $\Delta\tilde{s}$. When comparing the bending angles in the joint segments for a single bending shape in Fig. 4 it is noticeable that the angles are equal for the joint segments S_I and S_{II}. This observation is based on the fact, that the bending moment along the mechanism's longitudinal axis is constant when gravity is not considered. A more detailed consideration is given in Sect. 4.3.

Regarding the free end of the mechanism in Fig. 4, it is possible to achieve an orientation angle up to 270° with respect to the orientation of the free end in the undeformed configuration. Such bending shapes are well suited for example in endoscopic applications. It should be noted, that the model also delivers bending shapes for larger values of $\Delta\tilde{s}$. But it should be taken into account that the bending of the mechanism occurs in the plane. Therefore, the bending angle is limited due to possible collisions of the mechanism's free end with other mechanism segments.

In Fig. 5 the stiffness factors are changed to $c_I = 0.2$ and $c_{II} = 0.05$. In comparison to Fig. 4 it is obvious, that the joint angle in joint segment S_{II} increases while the angle in joint segment S_I decreases. Therefore, the bending of the finger-like mechanism is more focused on the fingertip. The orientation angle of the mechanism's free end are comparable to Fig. 4 but the bending shapes clearly differ.

The bending shapes shown in Fig. 4 and Fig. 5 are intended to exemplify the possibilities to set up a desired bending shape for the compliant mechanism only by adjusting the stiffness factors c_I and c_{II} of the joint segments. Of course there are other parameters that can be used to set up the bending shape, such as the length of the segments.

4.2 Bending Shapes Under Consideration of Gravity

As described in Sect. 3 the influence of gravity (weight of the mechanism) on the resulting bending shape can be considered by setting the parameter $\tilde{q}_w$ to values $\tilde{q}_w > 0$, see Eqs. (3) and (4). To simplify matters, it is assumed that the mass of the mechanism is equally distributed over its length l. Of course it is possible to calculate any desired distribution of mass, for example in case the weight of the joint segments differs from weight of the rigid segments. By adapting the model, it is also possible to calculate the bending shape taking a desired payload into account.

In Fig. 6 the resulting bending shapes under consideration of $\tilde{q}_w = 1$ are shown for travel distances of the cable $\Delta\tilde{s}$ between 0 and 0.22. The gravity works in negative direction of the y-axis. For comparison, the bending shapes from Fig. 4 (without consideration of weight) are additionally shown in Fig. 6. It is obvious that the position of the mechanism's free end is shifted in direction of gravity when weight is considered and small values of $\Delta\tilde{s}$ are regarded.

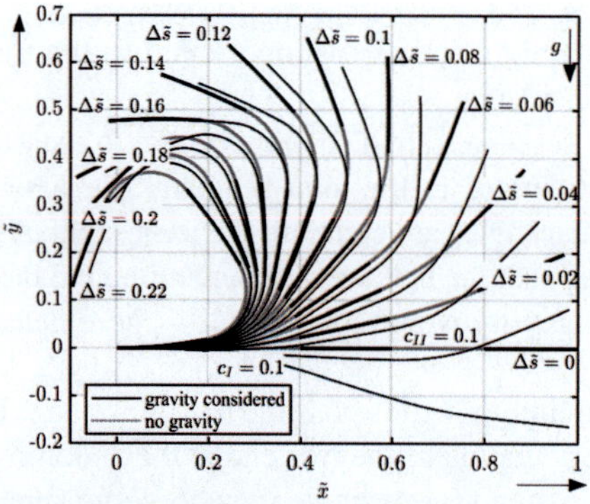

Fig. 6. Calculated bending shapes under consideration of gravity (negative y-direction)

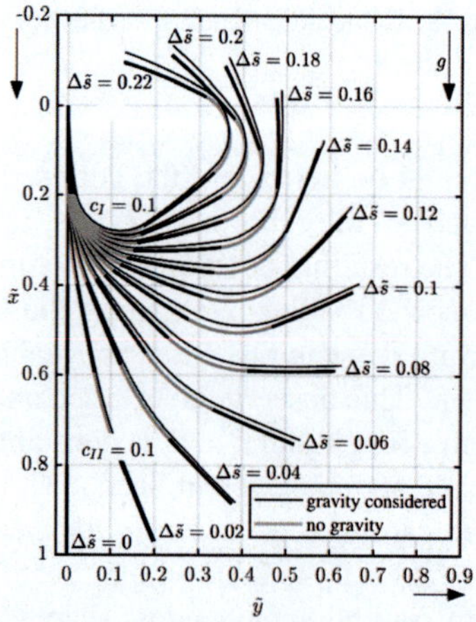

Fig. 7. Calculated bending shapes under consideration of gravity (positive x-direction)

In Fig. 7 the mechanism is shown in a 90° clockwise rotated configuration. The gravity works in positive direction of the x-axis. As can be seen, the influence of gravity is much smaller for this configuration. It is also interesting to observe that for some bending shapes in Fig. 7 the influence of gravity leads to a lifting of the mechanism's free end, thus changing its position against the direction of gravity.

It should be noted, that the results from Fig. 6 and Fig. 7 are not surprising. The intention is much more to show that the applied mechanical model from Sect. 3 for the description of the bending behavior leads to qualitatively correct results. In Sect. 5 the validity of the model is confirmed by experimental tests.

4.3 Bending Moment

Another important point to be discussed is the bending moment. It is obvious from Fig. 2 that the bending moment M on the mechanism's free end is calculated from the product of the lever arm a and the cable force F. It would normally be expected, that the bending moment is variable along the longitudinal axis of the mechanism, due to the differing distance of the force's acting point to the point on the mechanism, where M is currently considered. But the results from the calculations show that the bending moment is constant over the mechanism's length when gravity is not considered. This surprising insight is due to the line load $q_2(\kappa_3)$ which results from the cable guides (see Sect. 3).

For comparison, in Fig. 8 the resulting bending moments $\tilde{M}_3$ are exemplarily shown for the mechanism configuration from Fig. 6 under consideration of gravity and when gravity is not considered. As can be seen, the bending moment is not constant when gravity is considered. The resulting curvature of the mechanism is proportional to the bending moment.

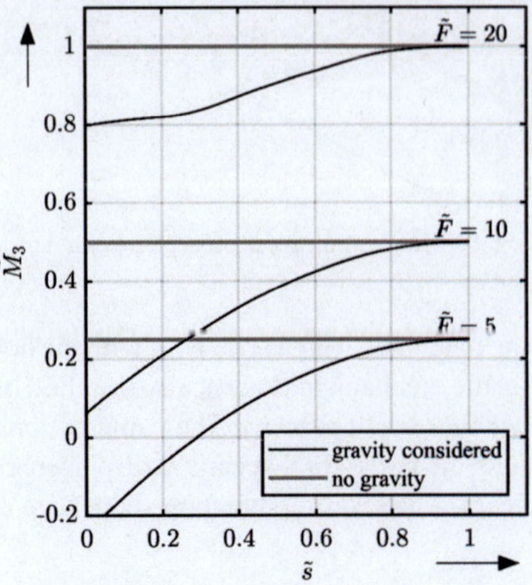

Fig. 8. Comparison of bending moment

It should be noted, that the resulting bending shapes of the investigated compliant mechanism in this paper only depend on the value of the bending moment that works at the free end of the mechanism. Thereby the composition of the bending moment M plays no role while the product of lever arm a and cable force F result in the same value.

5 Experimental Validation of the Model

The calculation model of the compliant mechanism presented in this paper uses a novel approach to modelling the actuation cable (see Sect. 3). In order to validate the model a prototype of the compliant finger-like mechanism is built up and tested. The aim is to compare the resulting bending shapes from the tested mechanism with those calculated from the mechanical model using the novel approach.

5.1 Experimental Setup and Test Procedure

The basic concept of the prototype mechanism for validation is shown in Fig. 9. The concept corresponds to the idea of the mechanism described in Sect. 2. The prototype mechanism is composed of three rigid segments (1) which are built up by 3D–printing (FDM) and two spring steel strips (2) which form the compliant joint segments. The cable (4) is guided by multiple 3D-printed cable guides (3). The free end of the mechanism is designed as an angled segment to allow attachment of the cable.

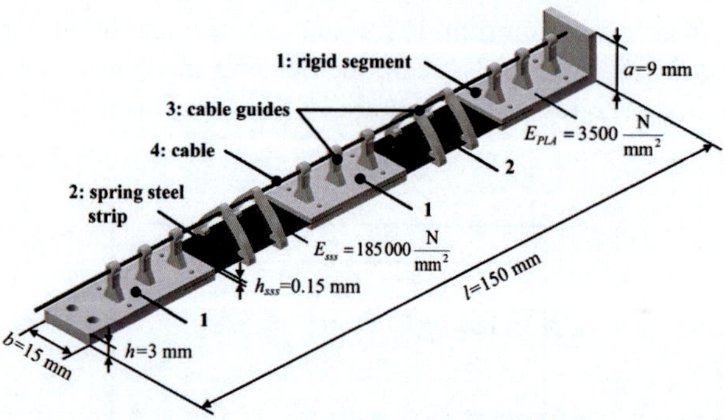

Fig. 9. Concept of the prototype mechanism

The experiments on the mechanism are carried out with the help of a universal testing machine. For this purpose, the mechanism is vertically installed in the machine as shown in Fig. 10 (the testing machine is not shown). The cable is connected to the universal testing machine which pulls the cable for a specified displacement Δs (travel distance) during test of the mechanism. The resulting bending shapes are captured with a camera for further evaluation.

5.2 Test Results

The resulting bending shapes of the prototype mechanism for different displacements Δs are shown in Fig. 10. For Δs values up to 30 mm are considered which corresponds to dimensionless values of $\Delta\tilde{s}$ up to 0.2.

In comparison to the calculated bending shapes shown in Fig. 7, a qualitative match of the experimentally determined shapes can be seen.

For the quantitative evaluation and the comparison with the calculated bending shapes (calculation model) the experimentally determined bending shapes (captured with a camera) are digitized by using the program *GetData Graph Digitizer*.

In the next step, the geometric dimensions of the prototype are converted into the corresponding dimensionless parameter values and fed into the calculation model. For the stiffness factor of the joint segments a value of $c_I = c_{II} = 0.0066$ is determined, for the lever arm a dimensionless value of $\tilde{a} = 0.06$ is found.

To compare the results from the experiments and calculations, the following assumption is made. If the model gives correct results, then the bending shapes from experiment and calculation should match. In addition, the values of $\Delta\tilde{s}$ should be the same for the considered bending shape. The evaluation is carried out by adjusting the value of $\Delta\tilde{s}$ in the calculation model until the resulting bending shape corresponds as closely as

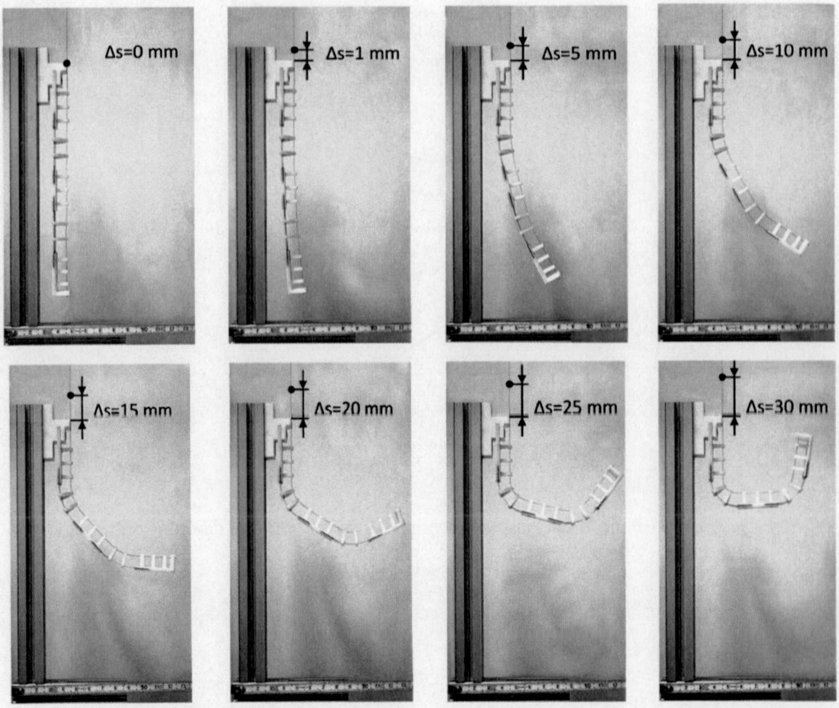

Fig. 10. Bending shapes (prototype) for different cable displacements Δs

possible with the digitized bending shapes from the experiment. In Fig. 11 the bending shapes from experiment and calculation are shown in superimposed representation (values of $\Delta\tilde{s}$ are given for the bending shapes from experiment). With the exception of the bending shape for $\Delta\tilde{s} = 0.2$ it can be stated that the bending shapes from experiment and calculation do not differ in practice.

In Table 1, the corresponding values of Δs from the calculation model and the experiments are shown. It should be noted that for presentation purposes, the values of $\Delta\tilde{s}$ from the calculation model are converted into dimensional values Δs. Table 1 shows that the absolute difference of Δs between experiment and calculation increases from 0.15 mm to 2.75 mm, when larger values of Δs are considered.

These results are very positive, since the calculation model does not fully correspond to the prototype. The prototype is equipped with punctual cable guides, while the model considers continuous cable guides. Thus, with increasing curvature of the mechanism, punctual guides lead to larger values of Δs as confirmed by experiment. Another point is the cable used in the experiments. The cable is not limp, but in the calculation model it is considered to be limp. Last but not least, inaccuracies in the digitalization of the bending forms which may result from the alignment of the camera cannot be ruled out.

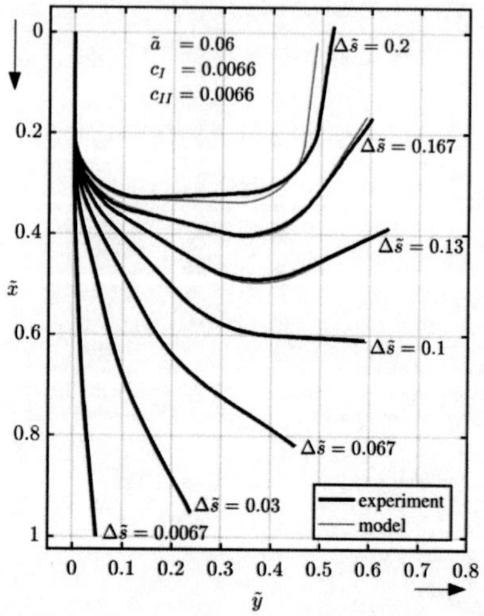

Fig. 11. Comparison of bending shapes from experiment and calculation model

Table 1. Comparison of Δs (displacement of the cable) from calculation and experiment

Δs [mm] (experiment)	Δs_{model} [mm] (calculation)	$\Delta s - \Delta s_{\text{model}}$ [mm]	Deviation [%]
1	0.86	0.15	14.5
5	4.35	0.65	13
10	8.99	1.02	10.15
15	13.71	1.29	8.6
20	18.26	1.75	8.73
25	23.1	1.9	7.6
30	27.26	2.75	9.15

6 Conclusions

In this paper, a novel approach to modeling a cable-driven finger-like compliant mechanism is presented. It is shown that on the basis of the model the bending forms of the mechanism can be calculated under consideration of large deformations. It is also possible to determine the influence of gravity and to consider payloads further. It is shown that the forces transmitted between the mechanism and the cable can be represented in the model by a line load which depends on the curvature of the mechanism. Finally, the model is confirmed by experiments.

The presented mechanism opens up a wide range of applications, based on the variety of resulting bending shapes, that can be adjusted by changing the bending stiffness in the joint segments. Further research should focus on the development and design of suitable joint segments with cable routing as a challenge. The calculation should be extended so that additional segments can be considered.

References

Catalano, M.G., Grioli, G., Farnioli, E., Serio, A., Piazza, C., Bicchi, A.: Adaptive synergies for the design and control of the Pisa/IIT SoftHand. Int. J. Robot. Res. **33**(5), 768–782 (2014). https://doi.org/10.1177/0278364913518998

Cheng, N.G., Lobovsky, M.B., Hosoi, A.E., Lagnemma, K.D.: Continuum style manipulator actuated with phase change media. Applied for by Massachusetts Institute of Technology (2013). Patent no. WO 2013 184192 A3

Deimel, R., Brock, O.: A novel type of compliant and underactuated robotic hand for dexterous grasping. Int. J. Robot. Res. **35**(13), 161–185 (2016). https://doi.org/10.1177/0278364915592961

Grebenstein, M., Chalon, M., Friedl, W., Haddadin, S., Wimböck, T., Hirzinger, G., Siegwart, R.: The hand of the DLR hand arm system: designed for interaction. Int. J. Robot. Res. **31**(13), 1531–1555 (2012). https://doi.org/10.1177/0278364912459209

Megaro, V., Knoop, E., Spielberg, A., Levin, D.I.W., Matusik, W., Gross, M., et al.: Designing cable-driven actuation networks for kinematic chains and trees. In: Teran, J. (ed.) Proceedings of the ACM SIGGRAPH Eurographics Symposium on Computer Animation. the ACM SIGGRAPH/Eurographics Symposium. Los Angeles, California, pp. 1–10. ACM, New York (2017)

Xu, Z., Todorov, E.: Design of a highly biomimetic anthropomorphic robotic hand towards artificial limb regeneration. In: Okamura, A., Menciassi, A. (eds.) 2016 IEEE International Conference on Robotics and Automation, Stockholm, Sweden, May 16–21, 2016 IEEE International Conference on Robotics and Automation (ICRA). Stockholm, Sweden. Institute of Electrical and Electronics Engineers; IEEE International Conference on Robotics and Automation; ICRA, pp. 3485–3492. IEEE, Piscataway (2016)

Zentner, L.: Nachgiebige Mechanismen. München: De Gruyter Oldenbourg (Studium) (2014)

Reconfigurable Planar Quadrilateral Linkages Based on the Tensegrity Principle

Philipp Schorr[1,2(✉)], Jhohan Chavez[1,2], Lena Zentner[1], and Valter Böhm[2]

[1] Technische Universität Ilmenau, Ilmenau, Germany
philipp.schorr@tu-ilmenau.de
[2] OTH Regensburg, Regensburg, Germany

Abstract. A feasible possibility to develop planar reconfigurable mechanisms is introduced in this work. Applying the tensegrity principle to common four-bar linkages allows a controllable change between two configurations of the mechanism. These two states correspond to different working spaces which vary regarding to the kinematic and mechanical properties. Therefore, the reconfiguration of the mechanism enables two different operation modes. Hence, this kind of mechanism enables the advantageous properties of conventional linkages with an additional enhanced adaptability of the kinematic and mechanic behavior. Beside the conceptual design of such tensegrity-based mechanisms, a reconfigurable four-bar parallel linkage is considered exemplarily. Numerical simulations are evaluated focusing on the kinematic behavior and the structural mechanics of this mechanism. Especially the reconfiguration of the mechanism by changing between two different working spaces is considered. The simulation results clarify the benefit of utilizing the tensegrity principle in mechanism theory. Adding only a few members to the original linkage enables a reconfigurable mechanism with comparable complexity.

Keywords: Tensegrity · Reconfigurable mechanisms · Planar four-bar linkage

1 Introduction

Often, in machines complex trajectories are required to realize sophisticated tasks. Although, mechanisms with multiple degrees of freedom, like robotic arms enable an immense adaptability to time-varying conditions, conventional linkages are still used because of the corresponding accuracy of the trajectory. However, such mechanisms are usually optimized for only one special task. Therefore, there is no possibility to tune the kinematic and mechanical behavior. This fact encourages the consideration of reconfigurable mechanisms which enable a change between different operation modes. In [1–3] the reconfiguration of the mechanisms is realized by using complex joints which enable different states corresponding to various kinematic properties. Various approaches to realize a reconfiguration are given in [4, 5]. However, all of these approaches yield an increase of the complexity of the mechanisms. Recently, the investigation of prestressed mechanisms, especially cable-driven mechanisms is focused by researchers.

L. Zentner and S. Strehle (Eds.): MAMM 2020, MMS 96, pp. 48–57, 2021.
https://doi.org/10.1007/978-3-030-61652-6_5

Those mechanisms are controlled by varying the prestress state and enable an adaptability of the mechanical properties [6]. However, the kinematic properties are limited [7]. The application of tensegrity structures in mechanism theory represents another approach [8, 9]. Although, those structures were established in the fields of architecture [10] and sculpturing [11] they feature promising properties for an use in engineering [12]. In particular, tensegrity structures enabling multiple stable equilibrium configurations, so-called multistable tensegrity structures, allow an adaptability of the mechanical and kinetic properties [13]. This fact is utilized in [14, 15] for the application as gripper.

In this work, an approach to realize reconfigurable mechanisms utilizing the tensegrity principle is presented. Following planar quadrilateral linkages are considered. The mechanical model of these mechanisms is discussed, and different working spaces are derived. In order to change between different operation modes, the application of ropes is suggested. However, the tension of these ropes must be guaranteed in operation. Thus, the linkages are extended to a tensegrity-based mechanism. This approach is presented exemplarily for a parallel linkage. The corresponding kinematic and mechanical properties are investigated in simulations. Especially, the change between the different working spaces of the mechanism is focused. The results confirm the reconfiguration and show the benefit by applying the tensegrity principle.

2 Planar Four-Bar Linkages

2.1 Structural Assembly and Dynamic Modeling

In this work, planar quadrilateral linkages with revolute joints are considered. The corresponding kinematic chains are closed, and the links are assumed as rigid. According to KUTZBACH–GRÜBLER's equation, this kind of mechanism features one degree of freedom. The current configuration of the four-bar linkage is described by the joint positions $\boldsymbol{r}_i = (x_i, y_i)^{\mathrm{T}} (i = 1, 2, 3, 4)$ in the x-y-plane as shown in Fig. 1. The enumeration of the joints and the links is chosen according to the conventional nomenclature of linkages. Thus, link 1 is fixed, the driving angle is named φ and the output angle is defined by ψ.

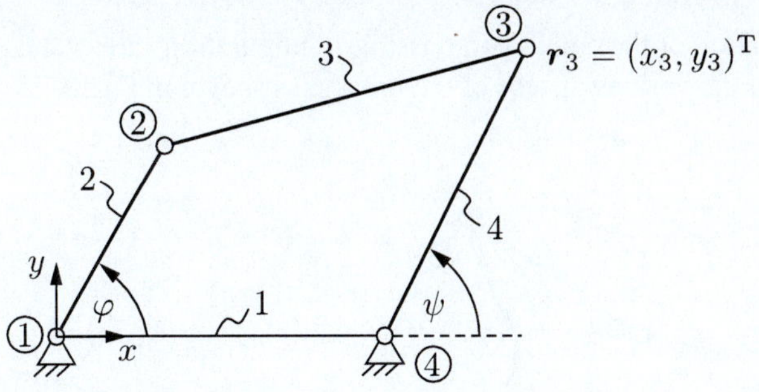

Fig. 1. Link and joint numeration of planar four-bar linkage in the x-y-plane.

To describe the kinematic and dynamic behavior of these mechanisms a mechanical model and the equations of motion are required. The links $(j = 1, 2, 3, 4)$ are defined by the constant length L_j, the mass m_j and the moment of inertia $\boldsymbol{\Theta}_j$. Friction in the joints is neglected. Due to the two-dimensional consideration and the given support, this approach yields 4 generalized coordinates summarized in $\boldsymbol{q} = (x_2, y_2, x_3, y_3)^{\mathrm{T}}$. Moreover, the holonomic constraints formulated in (1) must be taken into account.

$$g_b = 0 \text{ with } g_b = L_{b+1} - \sqrt{(x_{b+1} - x_b)^2 + (y_{b+1} - y_b)^2} \quad (b = 1, 2, 3) \tag{1}$$

The LAGRANGE formalism with multipliers λ is applied to derive the equations of motion. This approach yields the system of differential-algebraic equations shown in (2)–(3) where the parameters T and Q represent the kinetic energy and the generalized forces. To consider energy dissipation, STOKES friction is considered at the nodes (coefficient $c = 0.4$ Ns/m).

$$\frac{\mathrm{d}}{\mathrm{d}t}\frac{\partial T}{\partial \dot{q}_a} - \frac{\partial T}{\partial q_a} = Q_a + \lambda_b \frac{\partial g_b}{\partial q_a} \quad (a = 1, 2, 3, 4) \tag{2}$$

$$g_b = 0 \quad (b = 1, 2, 3) \tag{3}$$

These nonlinear equations of motion are solved numerically applying *Numerical Differentiation Formulas.* To avoid the drift-off of the holonomic constraints due to numerical accuracy the GEAR-GUPTA-LEIMKUHLER stabilization [16, 17] is applied.

2.2 Working Space

To determine the working space of the planar linkage the GRASHOF condition (4a) is considered. If this condition is fulfilled the link with minimum length is defined as crank and can rotate completely respective to the neighboring members.

$$l_{\min} + l_{\max} < l' + l'' \tag{4a}$$

$$l_{\min} + l_{\max} > l' + l'' \tag{4b}$$

$$l_{\min} + l_{\max} = l' + l'' \tag{4c}$$

For each orientation of the crank, two possible configurations are evaluated. This issue yields two separate working spaces of the linkage as shown in Fig. 2.

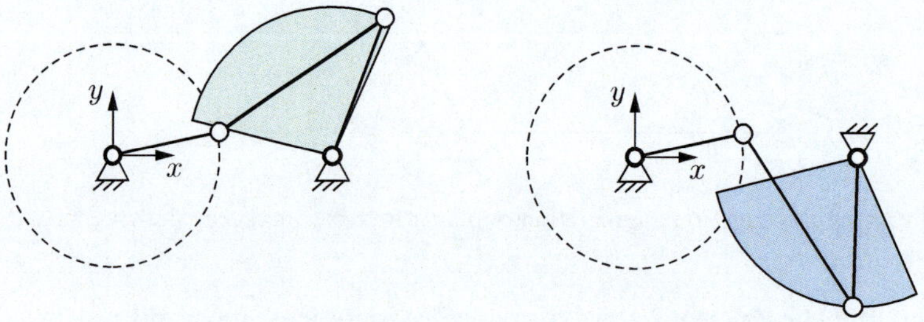

Fig. 2. Working spaces of planar four-bar linkage according to (4a).

These working spaces correspond to dissimilar configurations with different kinematic and mechanical properties. This fact is clarified considering the trajectory of a point attached to link 3 as shown in Fig. 3. Obviously, different tasks can be realized utilizing the various working spaces. This issue motivates the consideration of the change between these configurations, the so-called reconfiguration of the mechanism. However, in the considered linkages this change is not possible due to the holonomic constraints. Only disassembling the mechanisms allows switching the operation mode.

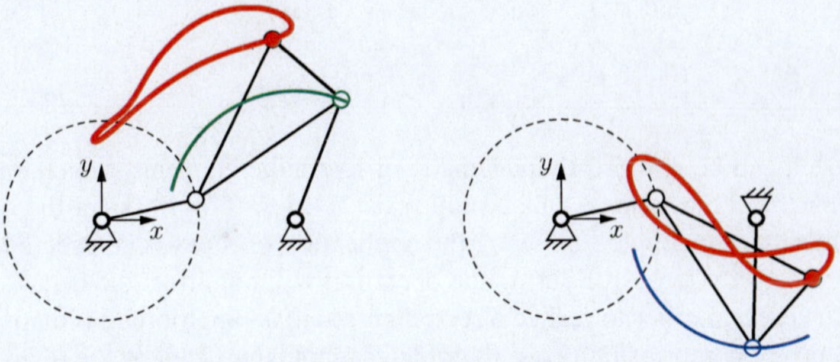

Fig. 3. Trajectories of arbitrary node attached to link 3 different working spaces.

Considering mechanisms fulfilling the condition (4b) also yields two separate working spaces (see Fig. 4 a). Only quadrilateral linkages characterized by (4c) enable an intersection of the two working spaces (see Fig. 4 b). However, according to [18] a controllable change between these operation modes is not possible. Thus, conventional quadrilateral linkages do not enable a reconfiguration.

a)

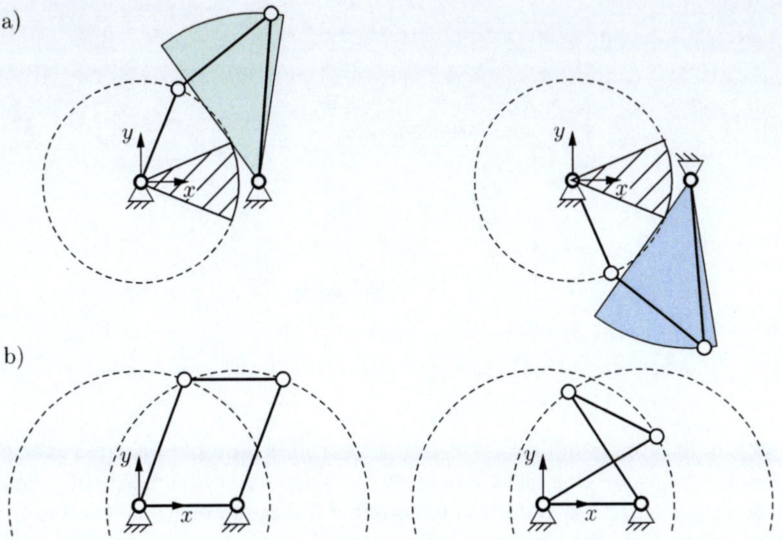

Fig. 4. Working spaces of linkages according to GRASHOF condition – a) (4b), b) (4c).

2.3 Reconfiguration of Quadrilateral Linkages

Obviously, the holonomic constraints of the mechanism prevent the change between the various configuration spaces. However, these constraints guarantee the accuracy of the trajectory and the kinematic behavior in general. In order to allow a reconfiguration of the mechanism without affecting these advantageous properties in operation the holonomic constraints of selected links are replaced by one-sided limited nonholonomic constraints as shown in (5a) or (5b).

$$g_c \geq 0 \quad (c \in [1, 2, 3]) \tag{5a}$$

$$g_c \leq 0 \quad (c \in [1, 2, 3]) \tag{5b}$$

This approach can be realized by pneumatic or hydraulic elements, which only resist compression (5a) or by ropes, which only resist tension (5b). To keep the complexity of the mechanism simple, following the application of ropes is considered. Hence, guaranteeing tension ($\lambda_c > 0$) during the operation will not affect the behavior of the linkage. However, in order to realize a reconfiguration, an appropriate actuation yields the occurrence of compression ($\lambda_c < 0$), which cannot be resisted by the ropes. Hence, an additional degree of freedom exists, which allows a reconfiguration of the linkage. This approach is depicted qualitatively in Fig. 4.

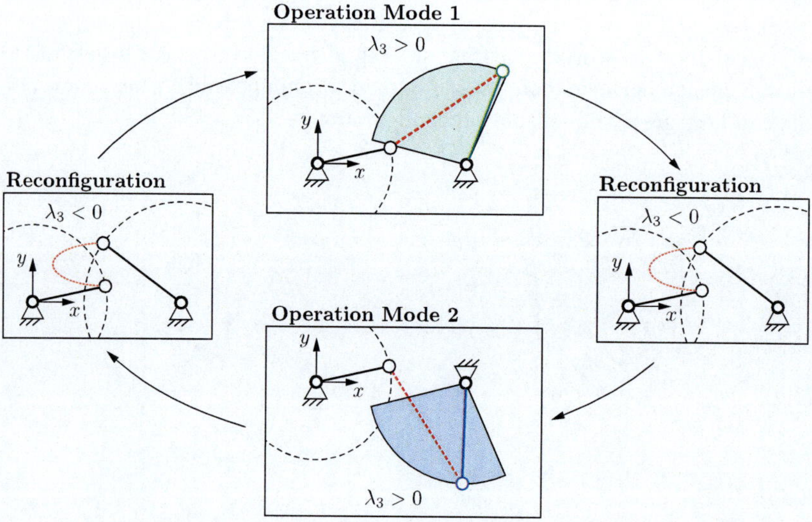

Fig. 5. Reconfiguration of the linkage by replacing link 3 with a rope according to (5b).

To ensure tension within the ropes during operation a proper prestress state of the mechanism is requires. This fact encourages the utilization of the tensegrity principle. Therefore, few segments are added to the original linkage to realize a tensegrity-based mechanism. Although, the kinematic behavior of the linkage in operation are identical, the mechanical properties change. Following, this issue is considered in detail.

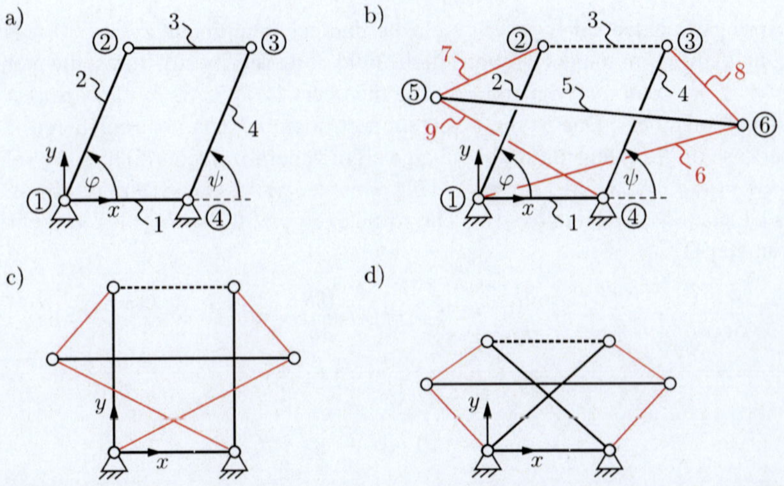

Fig. 6. Reconfiguration of a parallel linkage – a) original four-bar linkage, b) extension to tensegrity-based mechanism, c) stable equilibrium configuration corresponding to operation mode 1, d) stable equilibrium configuration corresponding to operation mode 2.

3 Tensegrity-Based Mechanism

3.1 Structural Topology

The discussed approach is considered exemplarily for the parallel linkage displayed in Fig. 6 a. This four-bar linkage also features two working spaces (see Fig. 4c). To enable a reconfiguration of the mechanism link 3 is replaced by a rope. Therefore, in operation this member must be loaded by tension. To ensure this load case one link (member 5) and four springs (members 6–9, shown in red) are added to the planar linkage to realize a tensegrity-based mechanism as shown in Fig. 6 b. The parameters of the mechanism are listed in Table 1. The parameters of the members 6–9 are orientated to real springs (Fig. 5).

Table 1. Applied mechanical parameters of the tensegrity-based parallel linkage.

	Conventional linkage				Extension to tensegrity-based mechanism				
j	1	2	3	4	5	6	7	8	9
l (m)	0.075	0.100	l_1	l_2	0.150	0.0249	0.0236	0.0236	0.0249
m (kg)	0.075	0.100	0	0.100	0.150	–	–	–	–
k (N/m)	–	–	–	–	–	72	128	128	72

To describe the structural dynamics, the mechanical model of the linkage is extended. The rope is supposed to be massless and is only considered by the nonholonomic constraint. The springs are specified by the stiffness k and the initial length l. To take energy dissipation due to the deformation into account, the mechanical model of the springs is extended by a visco-elastic damper (coefficient $c = 0.4$ Ns/m). According to the common declarations in the field of tensegrity structures, the members $j = 2, 4, 5$ are compressed members. The members $j = 3, 6, 7, 8, 9$ are classified as tensioned members. Due to the given support, member 1 is not considered. Moreover, because of the extension of the linkage eight generalized coordinates are selected $q = (x_2, y_2, x_3, y_3, x_5, y_5, x_6, y_6)^{\mathrm{T}}$. This approach yields the system of differential-algebraic equations shown in (6)–(8). The strain energy of the springs is represented by the parameter Π.

$$\frac{\mathrm{d}}{\mathrm{d}t}\frac{\partial T}{\partial \dot{q}_a} - \frac{\partial T}{\partial q_a} + \frac{\partial \Pi}{\partial q_a} = Q_a + \lambda_b \frac{\partial g_{b,c}}{\partial q_a} \quad (a = 1, 2, \ldots, 8) \tag{6}$$

$$g_b = 0 \quad (b = 1, 2, 4, 5) \tag{7}$$

$$g_c \leq 0 \quad (c = 3) \tag{8}$$

Solving these equations of motion numerically, two relevant stable equilibria, which correspond to the different working spaces, are detected (see Fig. 6 c and d).

3.2 Mechanical Behavior

In order to predict the actuation parameters of the tensegrity-based mechanism the static properties of the linkage are considered. A torque M_1 applied at node 1 is assumed as actuation. Due to the tension in rope 3 in operation, the holonomic constraints of the original linkage are fulfilled. Therefore, the kinematic transmission behavior of the tensegrity-based mechanism and the linkage is identical. The transmission behavior for the two working spaces are depicted in Fig. 7 a.

However, with regard to mechanical properties a great difference between the original linkage and the tensegrity-based mechanism exists. Due to the prestress stable equilibrium configurations exist. Without external actuation one of these states will occur. In contrast to the original linkage, triggering any other configuration requires an actuation torque. The torque-deflection-plot is shown in Fig. 7 b for the different working spaces. These results clarify the different static properties of the mechanisms. This fact must be considered for the design of the actuation strategy.

3.3 Reconfiguration of the Mechanism

To verify the discussed approach the reconfiguration of the planar tensegrity-based mechanism is considered exemplarily. During the reconfiguration the mechanism enables two degrees of freedom. Thus, in order to control the change between the different configurations two actuators are required. Beside the parameter M_1 a second torque M_2 is acting on node 5. In Fig. 8 the simulation results for changing from operation mode 1 to operation mode 2 are displayed for a given actuation strategy.

Indeed, the results confirm that the actuation (see Fig. 8 a) yields compression in rope 3. This member cannot resist that load and slackens (see Fig. 8 b). This fact initializes the reconfiguration of the mechanism. After reaching operation mode 2 the rope is again loaded by tension. Various states during the reconfiguration and trajectories of selected joints are depicted in Fig. 8 c. For some instances the tensioned members are compressed. Nevertheless, these simulation results show the benefit of the tensegrity principle and motivates further investigations.

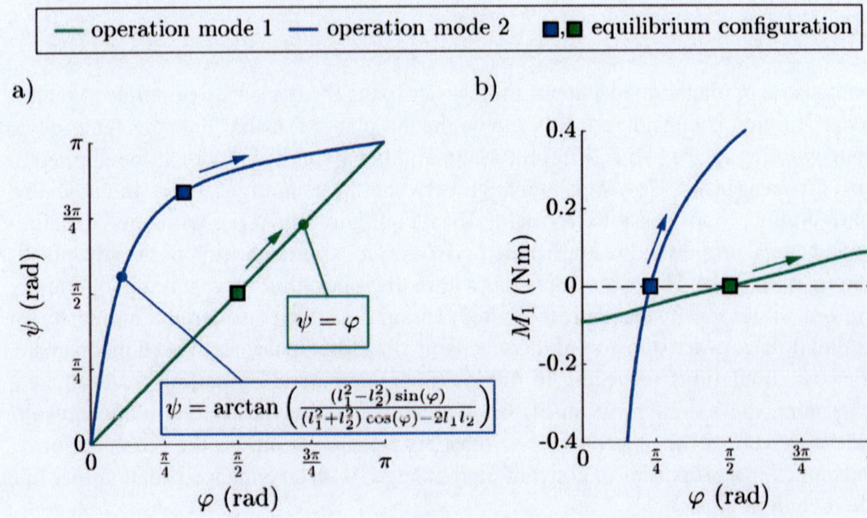

Fig. 7. Static investigation of the tensegrity-based mechanism in different operation modes according to Fig. 6 c and d – a) kinematic transmission behavior, b) static behavior.

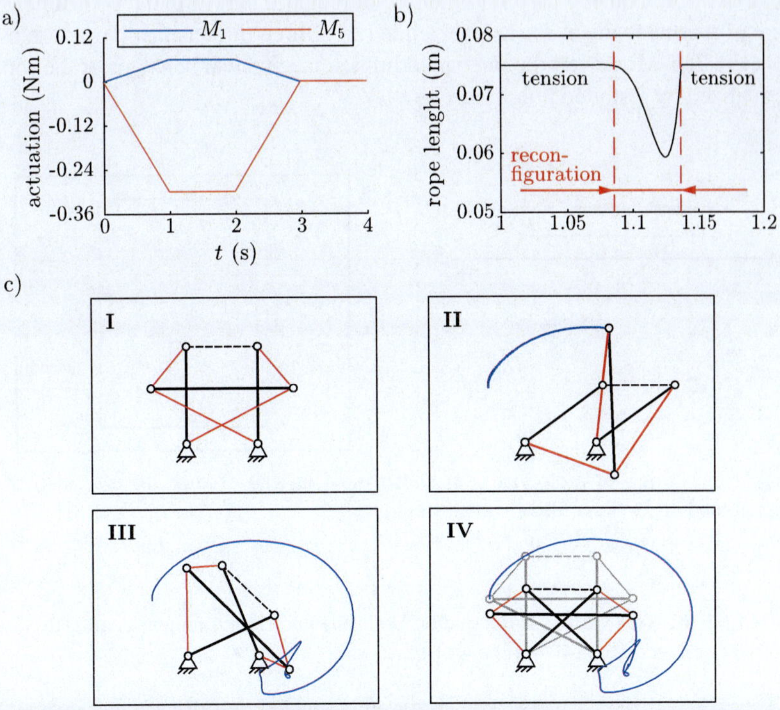

Fig. 8. Simulation of the reconfiguration of the tensegrity-based mechanism – a) applied actuation strategy, b) length of the rope during the reconfiguration, c) various states of the mechanism during the reconfiguration and trajectory of node 5.

4 Conclusion

The extension of planar quadrilateral linkages utilizing the tensegrity principle to realize a reconfiguration is considered. It is shown that all planar four-bar linkages feature two various working spaces with different kinematic behavior. This fact can be applied to realize different tasks. However, a change between these configurations, the so-called reconfiguration, is not possible. To realize the reconfiguration, ropes with one-sided limited mechanical properties are applied in this work. This approach remains the advantages of common linkages. However, a tension within the ropes must be ensured in operation. Therefore, the tensegrity principle is applied. This approach is considered exemplarily for a parallel linkage. An extension of the mechanism yields a tensegrity-based mechanism. Numerical simulations regarding to the kinematic and static properties are evaluated. Furthermore, the reconfiguration of the mechanism is shown. The simulation results show the benefit of this approach. The tensegrity principle allows the combination of the advantageous properties of conventional linkages with an enhanced adaptability due to the reconfiguration.

These results encourages experimental investigations of tensegrity-based mechanisms. In particular, the kinematic and mechanical properties as well as the reconfiguration of the mechanism should be verified in experiments. Thus, the development of a prototype of the considered parallel linkage is targeted. Therefore, the two-dimensional topology of the mechanism must be extended to a three-dimensional structure as presented in Fig. 9 b. Moreover, the one-sided limited mechanical behavior of the rope can also be realized by a slotted hole (see Fig. 9 a).

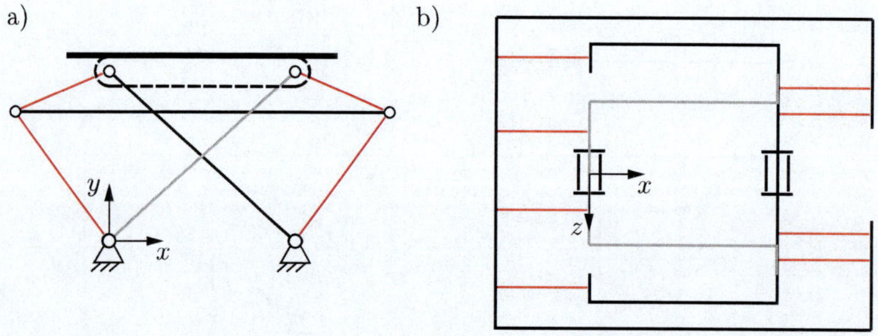

Fig. 9. Possible design of prototype – a) slotted hole instead of rope, b) extension of two-dimensional topology to three-dimensional structure.

Acknowledgment. This work is supported by Deutsche Forschungsgemeinschaft (DFG) within the SPP 2100 - projects ZE714/14-1, BO4114/3-1.

References

1. Carbonari, L., Corinaldi, D., Palmieri, G., Palpacelli, M.C.: Kinematics of a novel 3-URU reconfigurable parallel robot. In: 2018 International Conference on Reconfigurable Mechanisms and Robots (ReMAR), pp. 1–7 (2018)
2. Wei, G., Dai, J.S.: Reconfigurable and deployable platonic mechanisms with a variable revolute joint. In: Lenarčič, J., Khatib, O. (eds.) Advances in Robot Kinematics, pp. 485–495. Springer, Cham (2014)
3. Nurahmi, L., Gan, D.: Operation mode and workspace of a 3-rRPS metamorphic parallel mechanism with a reconfigurable revolute joint. In: 2018 International Conference on Reconfigurable Mechanisms and Robots (ReMAR), pp. 1–9 (2018)
4. Kuo, C.-H., Dai, J.S., Yan, H.-S.: Reconfiguration principles and strategies for reconfigurable mechanisms. In: 2009 ASME/IFToMM International Conference on Reconfigurable Mechanisms and Robots, pp. 1–7 (2009)
5. Aimedee, F., Gogu, G., Dai, J.S., Bouzgarrou, C., Bouton, N.: Systematization of morphing in reconfigurable mechanisms. Mech. Mach. Theory **96**, 215–224 (2016)
6. Merlet, J.-P.: Preliminaries of a new approach for the direct kinematics of suspended cable-driven parallel robot with deformable cables. In: Wenger, P., Flores, P. (eds.) New Trends in Mechanism and Machine Science. Mechanisms and Machine Science, vol. 43, pp. 355–362. Springer, Cham (2017)
7. Yuan, H., Li, Z.: Workspace analysis of cable-driven continuum manipulators based on static model. Robot. Comput.-Integr. Manuf. **49**, 240–252 (2018)
8. Boehler, Q., Abdelaziz, S., Vedrines, M., Poignet, P., Renaud, P.: From modeling to control of a variable stiffness device based on a cable-driven tensegrity mechanism. Mech. Mach. Theory **107**, 1–12 (2017)
9. Fasquelle, B., Furet, M., Chevallereau, C., Wenger, P.: Dynamic modeling and control of a tensegrity manipulator mimicking a bird neck. In: Uhl, T. (ed.) Advances in Mechanism and Machine Science. IFToMM WC 2019, pp. 2087–2097, Springer, Cham (2019)
10. Fuller, R.B.: Tensile-Integrity Structures. U.S. Patent Nr. 3,063,521 (1962)
11. Snelson, K.: Continuous Tension, Discontinuous Compression Structures. U.S. Patent Nr. 3,169,611 (1965)
12. Rieffel, J., Mouret, J.-B.: Adaptive and resilient soft tensegrity robots. Soft Robot. **5**(3), 318–329 (2018)
13. Xu, X., Luo, Y.: Multistable tensegrity structures. J. Struct. Eng. **137**(1), 117–123 (2010)
14. Sumi, S., Böhm, V., Zimmermann, K.: A multistable tensegrity structure with a gripper application. Mech. Mach. Theory **114**, 204–217 (2017)
15. Sumi, S., Böhm, V., Schorr, P., Zentner, L., Zimmermann, K.: Compliant class 1 tensegrity structures for gripper applications. In: Corves, B., Wenger, P., Hüsing, M. (eds.) EuCoMeS 2018, pp. 392–399. Springer, Cham (2019)
16. Pöll, C., Hafner, I.: Index reduction and regularisation methods for multibody systems. IFAC-PapersOnLine **48**(1), 306–311 (2015)
17. Koutsovasilis, P., Beitelschmidt, M.: Simulation of constrained mechanical systems. PAMM: Proc. Appl. Math. Mech. **7**(1), 4010041–4010042 (2007)
18. Liu, G.F., Wu, Y.L., Wu, X.Z., Kuen, Y.Y., Li, Z.X.: Analysis and control of redundant parallel manipulators. In: Proceedings of the IEEE International Conference on Robotics and Automation, ICRA 2001, vol. 4, pp. 3748–3754 (2001)

Parameter Study of Compliant Elements for a Bipedal Robot to Increase Its Walking Efficiency

Marten Zirkel[1](✉), Yinnan Luo[2], Ulrich J. Römer[2], Alexander Fidlin[2], and Lena Zentner[1]

[1] Compliant Systems Group Technische Universität Ilmenau, 98693 Ilmenau, Germany
{marten.zirkel,lena.zentner}@tu-ilmenau.de

[2] Institute of Engineering Mechanics, Karlsruhe Institute of Technology (KIT), 76131 Karlsruhe, Germany
{yinnan.luo,ulrich.roemer,alexander.fidlin}@kit.edu

Abstract. In this paper, we introduce a method to place compliant elements with parameters that can be adjusted during operation in the joints of the bipedal walker to improve its energy efficiency. The bipedal walking robot is modelled with five rigid segments and driven by electric motors in its revolute joints. Minimizing the energy consumption of locomotion is formulated as a numerical optimization problem. An Euler-Bernoulli beam is used to describe the nonlinear behavior, caused by large deflections, of a compliant element loaded with forces and moments. The static problem for the beam deflection for given boundary conditions is solved numerically. Four parameters defining either the undeformed geometry or the boundary conditions are varied to modify the torque that this compliant element exerts on two robot segments connected by a revolute joint. The torque-deflection dependence and its dependence on the four different parameters is approximated by simple ansatz functions via fitting. The fitted functions are then included in a numerical optimization problem to determine the optimal parameters of the compliant element and the corresponding energy optimal gait simultaneously. We evaluate the optimized energy efficiency at different walking speeds, where the robot has different optimal gaits or parameters of the compliant elements. Two kinds of elastic couplings are investigated: the elastic coupling between the robot's upper body and its thighs; or between the robot's thighs and shanks. These specific compliant elements show a negligible performance gain from nonlinearity due to the small active operating range of these joints. However, the practicability of the proposed method for combining the detailed, model-based description of manufacturable compliant elements and the optimization of the overall robot system to achieve maximum energy efficiency is successfully demonstrated.

Keywords: Compliant elements · Bipedal robot · Numerical optimization · Energy efficiency

L. Zentner and S. Strehle (Eds.): MAMM 2020, MMS 96, pp. 58–75, 2021.
https://doi.org/10.1007/978-3-030-61652-6_6

1 Introduction

One of the major challenges in the development of bipedal walking robots is to improve their energy efficiency, which significantly determines the walking distance before the carried battery runs out. In contrast to the approaches pursued by Seok et al. with the MIT Cheetah [1], namely the design of the actuators and gears, or by Nasiri with impulsive actuation and control systems [2], our strategy for improving energy efficiency consists in the optimal elastic coupling of the robot segments by means of compliant mechanisms that temporarily store and release energy during the walking movement. This changes the natural dynamics during walking in such a way that the actuator power required to generate and stabilize a walking movement is reduced [3–6]. Through this approach, the efficiency of walking on level ground can be significantly improved just by using simple torsion springs with constant stiffness [7, 8]. However, this approach results in a specialization of the robot for the considered environmental conditions; a decrease in efficiency when walking in other environments (e.g. inclines, stairs) is expected. To overcome this drawback and reduce specialization, we propose the consideration of a large number of different environments in the optimization for future work. This approach, however, is restricted by the possibilities for designing and optimizing the compliant mechanisms. The consideration of merely a single parameter such as the constant torsional spring stiffness in earlier work severely limits the possible solution space. In this contribution, we investigate the possibilities of using nonlinear mechanisms to allow better adaptation to a larger number of operating conditions and thus counteract excessive specialization.

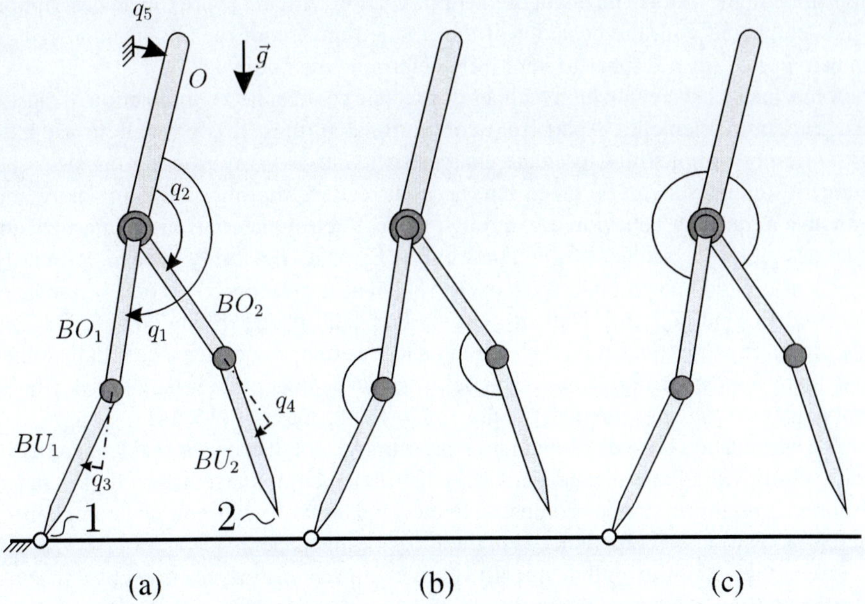

Fig. 1. (a) Model of the bipedal walker. (b) Compliant elements in the knees coupling the shanks with the thighs and (c) compliant elements in the hip connecting the shanks with the upper body.

The robot's energy efficiency depends not only on its mechanical design, but also on the control strategy that is used to generate a stable motion [9, 10]. In this first step, we limit our research on these influencing factors with an underactuated robot model, which consists of five rigid segments that are connected by compliant elements (CEs), see Fig. 1. To fully exploit the robot's natural dynamics, which can be modified through the characteristics of those compliant elements, we determine their design and parameters as well as the required walking movement simultaneously by numerical optimization [11]. The resulting optimal solution depends on the walking gaits as well as the environmental conditions. To extend the operating states of the robot, which benefit from the optimized compliant elements, a number of different ground inclinations is considered as opposed to just level ground. Even in fully actuated bipedal robots, CEs have an impact on the gaits, cf. [12].

In order to further improve the system's efficiency as compared to the preliminary work, we propose the use of CEs instead of simple torsion springs with constant stiffness as elastic couplings between the sections of the robot's body. With an optimized nonlinearity, the robot is able to adjust its natural dynamics for various walking conditions and can therefore improve its overall energy efficiency. Since the complete problem – the simultaneous optimization of the nonlinear CEs and the movement of the robot for many environments – is very complex, a two-step approach is taken: first, design solutions for CEs are developed and investigated. From these investigations, approximate descriptions for the behavior and characteristics are then determined as functions of a small number of parameters. This procedure should lead to a detailed and correct description of the CEs and the use of systematic methods for their design. At the same time, the approximate description of only a few parameters can be used to optimize them for the application in the robot. The result of the optimization with the approximate description is then compared with the behavior of the detailed models of the CE. This procedure can be repeated several times to iteratively determine the optimal CEs.

A mechanical system, which realizes desired curves, can be a combination of elastic and viscoelastic elements. Apart from conventional springs, it is possible to use CEs with different complexities to create characteristic curves. Any external physical and geometric influences can be taken into account to raise the physical complexity and to archive a transient behavior, e.g. through piezo-electric materials in the mechanism. If the design of the mechanism includes only CEs or its flexibility depends primarily on CEs, it is called a compliant system [13, 14]. These systems have some advantages compared to rigid body mechanisms. On the one hand, there are low to zero friction, easy processing due to a simple assembling and it is free of maintenance. On the other hand, there is mostly a highly nonlinear deformation behavior that is hard to describe in theory and the fatigue of the material due to large deformations [15, 16].

The compliance can be divided into concentrated and distributed due its geometric propagations. Compliant mechanisms with distributed compliance have a higher range of motion with lower stresses compared to mechanisms with concentrated compliance. Therefore, distributed compliance is in the focus of this paper.

Due to the possibility of dynamically changing the characteristic curves by adjusting an external influence during the operation of a compliant system in the walker, its gaits will also change and thus the potential should be there to expand the operating conditions of the walker.

Combined with large deflections, a well-designed strategy is required to solve this problem. For this task, an Euler-Bernoulli beam is described analytically and solved numerically in Matlab. We start our investigation with two static and two dynamically changeable geometric influences of a CE.

2 A Model of a Compliant Element

In this chapter, we introduce a CE in the form of a semicircle. This is a simple mechanism that nevertheless allows the connection of two rigid bodies parallel to a joint. To determine its characteristic curve – the torque-deflection dependency – one endpoint is moved on a constant radius around the joint by up to 130° in both directions. That implies large deformations. The basis of the model is an Euler-Bernoulli beam, which is introduced in the next section. Boundary conditions must be fulfilled at both endpoints of the beam, which is accomplished by a shooting method for the numerical solution. We vary four parameters that influence the boundary conditions and also the geometry of the mounted and of the dismounted beam. The behavior can be descripted with a single function which depends on these four parameters. This function is then fitted to the curves of the torque over the deflection as detailed below.

2.1 Base Model

Figure 2 shows a deflected beam with a rectangular cross section, length l, height h, width w, curvature κ_0 (which describes the curvature of the undeflected beam) and Young's modulus E. The length should be much larger than its height and width. In addition, we assume that its cross-sections do not deform and always stay perpendicular to the neutral axis. The material is assumed to be linear elastic and isotropic.

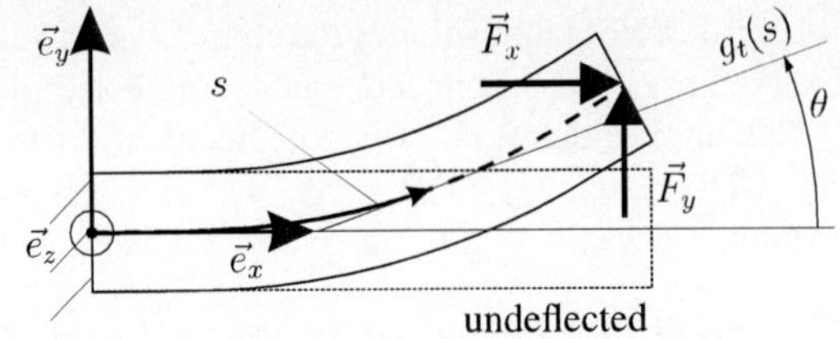

Fig. 2. External forces at the end of a beam cause bending deflection. The dotted lines show the natural form of the beam.

The height is measured in $\vec{e}_y$- and the width in $\vec{e}_z$-direction. In this model, the left hand side is clamped while the right hand side is loaded. The external forces $\vec{F}_x$ and $\vec{F}_y$ at $s = l$ of the neutral axis cause the deflection. It is also possible to load the end with an external torque M_{ext}. The dashed line depicts the neutral axis of this beam. θ specifies the angle between $\vec{e}_x$ and $g_t(s)$ which is the tangent of the neutral axis at s.

With this, it is possible to derive a system of ordinary differential equations (ODE) of the first order

$$\frac{dM}{ds} = F_x \sin\theta - F_y \cos\theta, \tag{1}$$

$$\frac{d\theta}{ds} = \frac{M}{EI_z} + \kappa_0 \; with \; I_z = \frac{wh^3}{12}, \tag{2}$$

$$\frac{dx}{ds} = \cos\theta, \tag{3}$$

$$\frac{dy}{ds} = \sin\theta, \tag{4}$$

as described in [17–19]. Given the external forces and the torques, this ODE can be solved numerically, for which we use a Runge-Kutta method. The geometry of the undeflected model can be specified as desired. In the next section, this model will be extended to fit the curve of a semicircle.

2.2 Semicircle

In Fig. 3, a CE in the shape of a semicircle that connects two links which are also connected by a planar revolute joint is displayed. The lower end of link 2 is clamped and link 1 can rotate around the joint. This forces the upper end of the element to move on a circle with a constant radius. The right side of Fig. 3 illustrates the functional sketch in which the compliant element is reduced to its neutral axis and the links are ideally rigid. So one end of the element is clamped and the other end can rotate around B_0. B indicates the undeflected state and in B' the end of the element connected to link 1 is moved by φ. In addition to the Cartesian coordinate frame, a cylindrical system that is fixed on the origin state of link 1 is introduced in B' to describe the external forces by a radial and a tangential component. As in Fig. 2, θ is the angle between the tangent $g_t(s)$ and $\vec{e}_x$ at s. The variable ρ_k describes the radius of the element in its desired position and therefore the desired curvature $\kappa_d = \frac{1}{\rho_k}$ in its mounted but undeflected state, which may be different from κ_0. This implies that if $\kappa_0 \equiv \kappa_d$, the beam is produced in the desired semicircle. To solve the ODE system (1)–(4), it is necessary to express the forces in the Cartesian coordinate frame

$$\frac{dM}{ds} = [\sin(\theta) - \cos(\theta)] \begin{bmatrix} \cos\varphi & -\sin\varphi \\ \sin\varphi & \cos\varphi \end{bmatrix} \begin{bmatrix} F_r \\ F_t \end{bmatrix}. \tag{5}$$

Furthermore, boundary conditions are required to determine the deflection of the compliant element. For A they are:

$$M(0) = M_{az}, \tag{6}$$

$$\theta(0) = \frac{\pi}{2}, \tag{7}$$

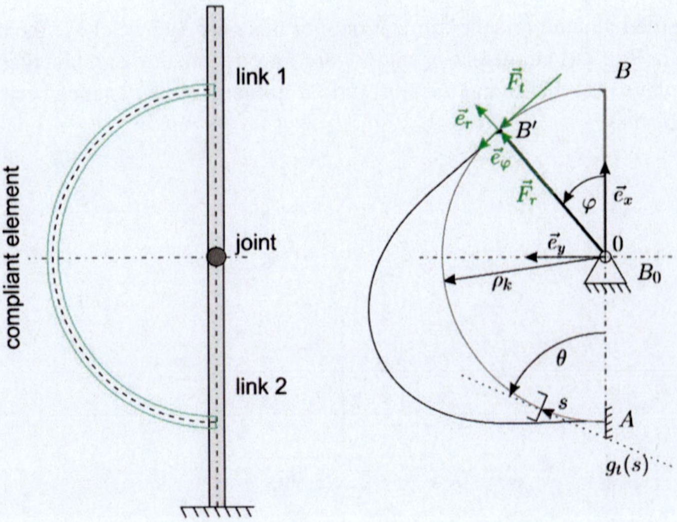

Fig. 3. Model of two links coupled over a joint by a compliant element on the left; and the corresponding functional sketch on the right.

$$x(0) = -\rho_k, \tag{8}$$

$$y(0) = 0. \tag{9}$$

With these boundary conditions, Eqs. (2)–(5) are solved by using *ode45* in Matlab. Since M_{az}, F_r and F_t are unknown, boundary conditions at B' are required to solve the system:

$$\theta(l) = -\frac{\pi}{2} + \varphi, \tag{10}$$

$$x(l) = \overline{B_0 B'} \cos\varphi, \tag{11}$$

$$y(l) = \overline{B_0 B'} \sin\varphi. \tag{12}$$

With an initial guess for M_{az}, F_t and F_r, the function f solve in Matlab is used to determine values for these variables, so that the solution of the ODE system satisfies the boundary conditions in B'. Notice that the torque $M(l)$ is not required for the solution. It can be set to an external torque M_{ext} and this comes with an extra initial guess of one of the boundary conditions.

The result includes the torque $M(l)$ at B'. With F_t, ρ_k and this torque, it is possible to calculate the torque M_{0z} in B_0

$$M_{0z} = F_t \rho_k + M(l). \tag{13}$$

The semicircular beam as introduced in this section is used as CE to couple the robot's segments as described in Sect. 1. Its parameters together with the corresponding gaits are to be determined via numerical optimization. For this purpose, we chose the parameters highlighted in Fig. 4. Parameters ρ_k and κ_d are introduced above; η changes the angle of the clampings in A and B and the shift ratio a influences the distance between B and B_0. It is defined as

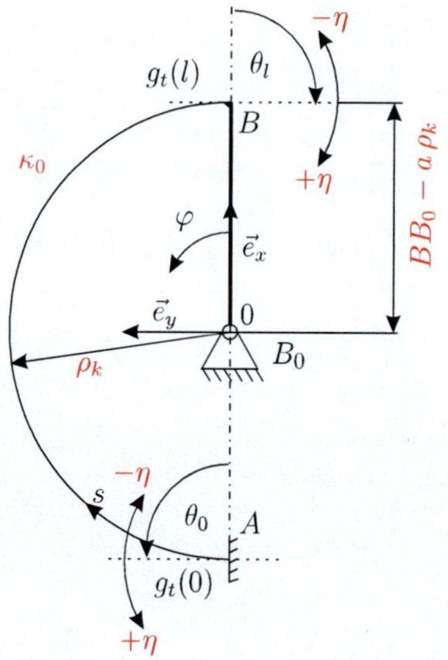

Fig. 4. Functional sketch of the semicircle with changeable parameters highlighted in red.

$$a = \frac{(\overline{BB}_0 - \rho_k)}{\rho_k}.$$

Remember that κ_0 describes the curvature of the beam in its natural form and κ_d describes the desired curvature.

3 Evaluation and Fitting

With the method introduced in the last section, we get the characteristic curve of the torque-degree dependency by solving the ODE system (2)–(5) with respect to the boundary conditions (6)–(12) and the parameter presented in the last section. Every parameter and variable, which describes a distance, a force or a torque is normalized by the height h or both the height and Young's modulus. That makes the mechanisms comparable and the material can be selected afterwards. The results are used to fit a function with 44 coefficients. These coefficients depend on the four variable parameter ρ_k, a, κ_0 and η. The function represents the CE in the joints of the bipedal walker shown in Fig. 1. With the method described in Sect. 3.2 we get the optimal values for ρ_k and κ_0 as well as a and η with respect to the walking speed.

3.1 Fitting

Before the element is investigated, a normalization with the height and Young's modulus is performed. This provides the possibility to compare elements with different geometries and makes the results independent from the material. Each geometry related parameter p_{geo} is normalized to the height and the forces as well as the torques are normalized to a product of the height and Young's modulus:

$$\tilde{p}_{geo} = \frac{p_{geo}}{h}, \tilde{F} = \frac{F}{Eh^2}, \tilde{M} = \frac{M}{Eh^3}.$$

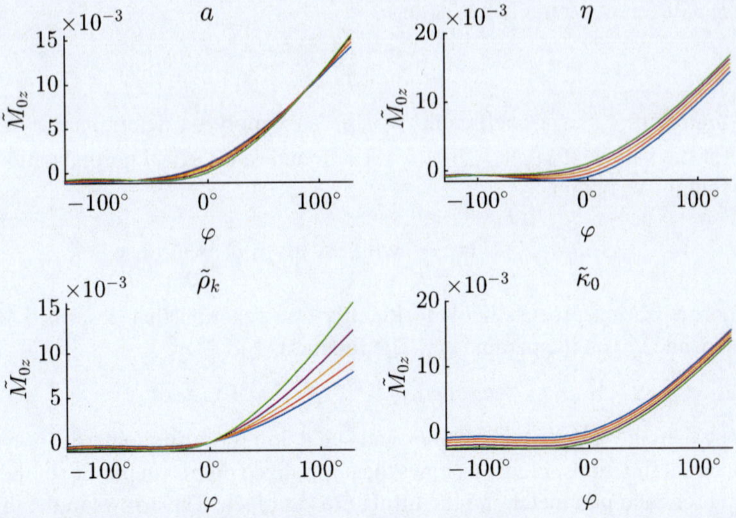

Fig. 5. Qualitative curves for the dependency of the torque $\tilde{M}_{0z}$ on variations of a (top left), η (top right), $\tilde{\rho}_k$ (bottom left) or $\tilde{\kappa}_0$ (bottom right).

The element is investigated as introduced in Sect. 2.2. Link 1 is rotated by angles between $\varphi \in [-130°, 130°]$. Also, each of the four chosen parameters has a standard value

$$a = 0, \eta = 0, \tilde{\rho}_\mathrm{k} = 50, \tilde{\kappa}_0 = -\tilde{\kappa}_d$$

and for the qualitative results in Fig. 5 only one of the parameters is changed at a time with five equidistant steps in the range of

$$a = [-0.1, 0.5], \eta = \left[-5°, 15°\right], \tilde{\rho}_k = [25, 50], \tilde{\kappa}_0 = \left[-\tilde{\kappa}_d, 0\right]. \tag{14}$$

The resulting curves for a, $\tilde{\rho}_k$ and $\tilde{\kappa}_0$ are normalized to their maximum value, so that $\tilde{M}_{0z}(\varphi = 130°) = 1$:

$$f_i = \tilde{M}_{0z}\left(\varphi = 130°, i\right) \text{with } i \in \{a, \eta, \tilde{\rho}_k, \tilde{\kappa}_0\}$$

For $\tilde{\kappa}_0$, an offset moment $\tilde{M}_{\kappa 0}$ of

$$M = (\kappa_d - \kappa_0)EI_z \tag{15}$$

$$\tilde{M}_{\kappa 0} = \frac{M}{Eh^3} = \frac{(\kappa_d - \kappa_0)I_z}{h^3} \tag{16}$$

is added to the values before the normalization is performed. If the natural curvature differs from the desired one, the beam must be bent. Equation (15) shows the relation between the difference of the curvatures and the torque which is required at the end of the beam. With (13) we get

$$\tilde{M}_{0z} = \tilde{F}_t \tilde{\rho}_k + \tilde{M}(l) + \tilde{M}_{\kappa 0}$$

For $\tilde{\rho}_k$, a different normalizing factor of

$$f_{\rho_k} = \frac{c_\rho}{\tilde{\rho}_k} \tag{17}$$

can be found with c_ρ as a coefficient, which can simply be determined by a fitting function for the values $\tilde{M}_{0z}(\varphi = 130^\circ, \tilde{\rho}_k)$. For a and η, no trivial normalization factor could be found. We obtain

$$\tilde{M}_{0z,norm,i} = \frac{\tilde{M}_{0z}}{f_i} \text{with } i \in \{a, \eta, \tilde{\rho}_k, \tilde{\kappa}_0\}.$$

The normalization process helps to identify one function that describes $\tilde{M}_{0z,norm}$ without dependency on the parameters. The function

$$\tilde{M} = c_1 - c_2 \cos(c_3\varphi)e^{c_4\varphi} + (c_5\varphi + c_6)e^{c_7\varphi} \tag{18}$$

fits best between the tested combinations and standalones of trigonometric, polynomial and exponential functions. Figure 6 shows the normalized points (in purple) of the torque $\tilde{M}_{0z}$ over φ for each parameter and the fitting (18) in black. The arrows in the third and fourth diagrams indicate an increase of the parameters η and a for the normalized points, respectively. Although the function (18) fits well for all four parameters, it lacks accuracy for a and η because of the simple normalization. To raise the accuracy of the function, the coefficients have to be modified in a way such that they are depending on the parameters η and a, because these are the parameters which have a nontrivial impact on the curve with respect to φ. Therefore, each coefficient is approximated by a polynomial of the second order as follows:

$$c_i(\eta, a) = c_{i1}a^2 + c_{i2}\eta^2 + c_{i3}\eta a + c_{i4}\eta + c_{i5}a + c_{i6}.$$

Additionally, with (16) and (17), Eq. (18) becomes

$$\tilde{M} = \frac{c_\rho}{f_{\rho_k}}[c_{\kappa 0}\tilde{M}_{\kappa 0} + c_1 - c_2 \cos(c_3\varphi)e^{c_4\varphi} + (c_5\varphi + c_6)e^{c_7\varphi}]. \tag{19}$$

Overall, there are 44 coefficients that must be determined, including c_ρ and $c_{\kappa 0}$. For this problem, the least squares of the torque $\tilde{M}_{0z}$ from (13) and $\tilde{M}(\rho, \kappa_0, \eta, a, \varphi)$ is build. The function lsqnonlin in Matlab varies the coefficients to minimize

$$\sum\nolimits_{(\rho_k, a, \eta, \kappa_0)} (\tilde{M}_0 z - \tilde{M})^2 \rightarrow min.$$

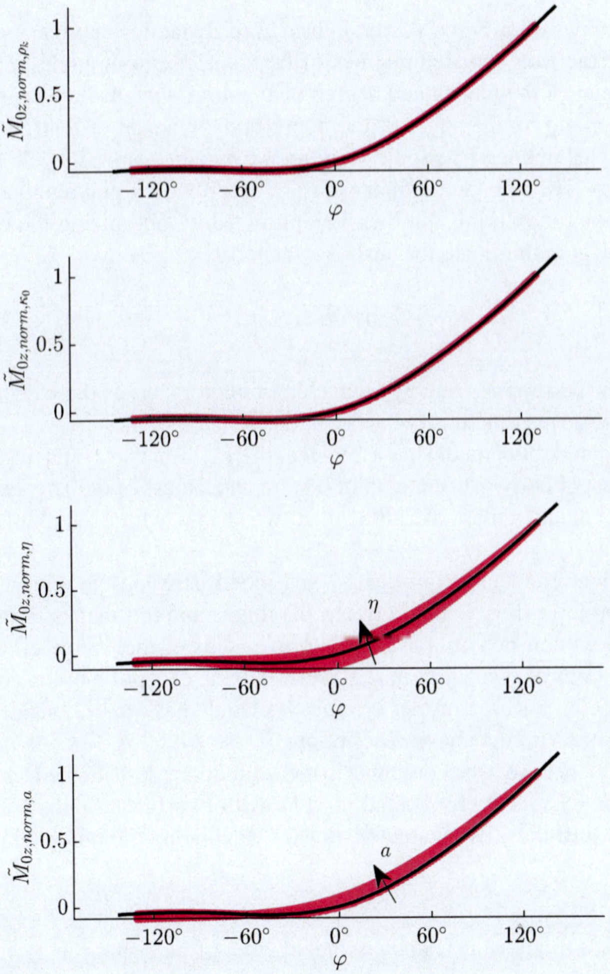

Fig. 6. Characteristic curves in (grey/purple) and the fitted function in black of the normalized torques depending on the four parameters from top to down: radius $\tilde{\rho}_k$, curvature $\tilde{\kappa}_0$, clamping degree η and shift ratio a.

3.2 Application of CEs in the Robot Model

As depicted in Fig. 1a, a planar robot model that consists of five rigid segments representing upper body, thighs and shanks is considered. Four electric motors are assumed to actuate the four ideal revolute joints in the hip and knees. The ends of the shanks are modeled as point feet so no torques can be transmitted between the robot's stance leg and the ground. Walking without slipping is achieved by additional constraints for unilateral contacts and static friction. We consider periodic walking gaits which consist of alternating single support phases, where only the stance leg is in contact with the ground, and instantaneous double support phases, where the former swing leg impacts the ground and the former stance leg lifts off without interaction. The gaits are produced and stabilized by a nonlinear controller based on the hybrid zero dynamics approach [7]. Reference trajectories for the four actuated degrees of freedom, the joint angles, are prescribed. These also influence the unactuated degree of freedom: the absolute orientation of the robot; e.g. described by the angle of the upper body. One set of reference trajectories for all joint angles defines a periodic walking movement – a limit cycle of this hybrid dynamic system. The reference trajectories are described by polynomials. The robot's energy efficiency for walking can then be optimized by optimizing the coefficients of these polynomials to minimize the cost of transport

$$COT = \frac{E_{electric}}{mgL}.$$

Here, $E_{electric}$ is the supplied energy from electric motors, mg is the weight force of the robot, and L is the step length.

The investigated robot model has a total length of 1.15 m and a total mass of 12,00 kg, containing an upper body with a length of 0.55 m and a mass of 6.00 kg, two thighs with a length of 0.30 m and a mass of 2.00 kg and two shanks with a length of 0.30 m and a mass of 1.00 kg.

The compliant elements from Sect. 2 are added between the upper body and the thighs in the hip joint (Fig. 1c) or between the thighs and the shanks in the knee joints (Fig. 1b) to study their potential use for improvements of the energy efficiency. These are all combinations of two adjacent segments with the exception of the coupling of the two thighs, since in this case further symmetries would have to be considered, which is not easily possible with the above description.

As the next step, the elastic couplings in the simulation are described by the nonlinear behaviors of the CEs as approximated in (17), with two further parameters k and φ_0. They scale and shift the CEs' characteristics for calculating the proper elastic torques:

$$M = k\tilde{M}\left(\tilde{\rho}_k, a, \eta, \tilde{\kappa}_0, \varphi - \varphi_0\right) \text{ with } k = Eh^3.$$

If the Young's modulus E is known, the height h of the beam can be calculated.

The robot's energy efficiency is evaluated for different walking speeds from 0.2 m/s to 1.4 m/s. We assume that after manufacturing, the parameters ρ_k, κ_0, k and φ_0 are constant while the parameters a and η can be still changed during walking movements according to the current operating state. We compare our results to ideal linear torsion springs to couple the robot's segments as described in [7], where two parameters (stiffness k and preload φ_0) define the linear behavior of a torsion spring. Since the stiffness k of the torsion spring is constant after it has been manufactured, we assume that the preload φ_0 can be individually adjusted for different walking speeds.

3.3 Workflow

The two separate methods from the previous subsections, the approximate description of the CE in Sect. 3.1 and the optimization of the robot in Sect. 3.2 are now combined to optimize the CE and the robot's gait simultaneously. The procedure is depicted in the flowchart in Fig. 7. The approximate description of the CE's characteristic curve and its dependency on the four varied parameters is calculated for an initial range for each parameter. This functional dependency of the moment on the angle and the parameters is then used in the numerical optimization procedure adapted from previous works. The parameters are constrained to the range which was used in the fitting procedure to get the approximate description. This measure is introduced to prevent large deviations between the approximate solution and the real solution, which are to be expected the further the range is left. If the optimum is out of the limits of the constrained parameter ranges, the range is adjusted and the fitting is performed again. This procedure is repeated iteratively until no further change in the parameter ranges is necessary or possible.

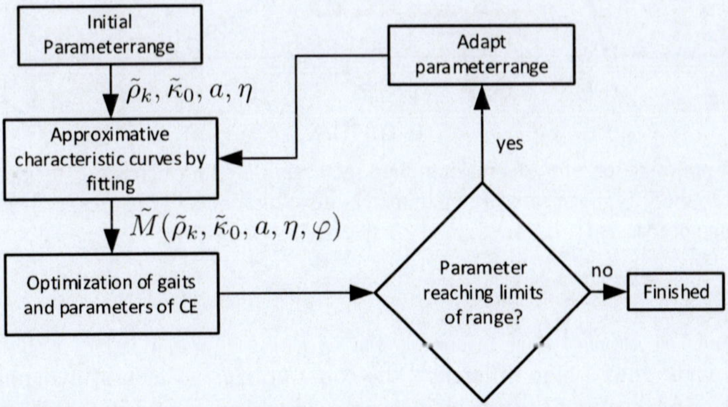

Fig. 7. The flowchart figures the process to optimize the gaits of the bipedal walker and the parameters of the CE.

The first iteration is initialized with the parameter ranges suggested in (14). As depicted in Fig. 8, the optimization of parameter a reaches the limit of -0.1 at walking speeds from 0.5 m/s–0.9 m/s; η reaches the limit of $-5°$ for walking speeds of 0.2 m/s–0.4 m/s and 1.3 m/s. Therefore, the ranges are adjusted to

$$a = [-0.3, 0.5],\ \eta = \left[-15°, 15°\right],$$

the results for which are discussed in the next section. It must be noted, that there are combinations of a and η in this range for which no solution can be found with the method mentioned in Sect. 3.1. To exclude this invalid parameter region, an additional linear constraint is added:

$$a \geq \frac{0.3}{15}\eta - 0.3.$$

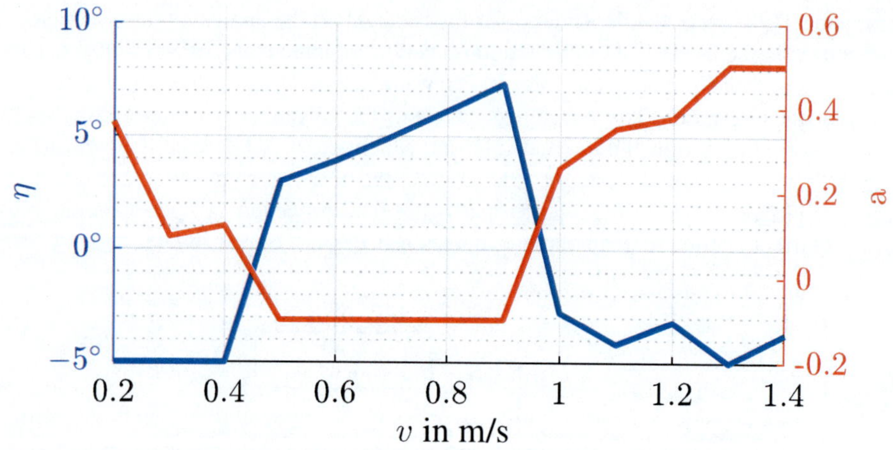

Fig. 8. First optimization of a (decreasing, then increasing) and η (increasing, then decreasing) with respect to the walking speed. Both parameters lie at the limits of their respective ranges for some walking speeds.

3.4 Results

In order to walk with a high efficiency, the robot utilizes its natural dynamics which mainly depend on the robot's mechanical design parameters, including the characteristics of the involved elastic couplings. Via numerical optimization, the robot's walking gaits as well as the parameters of the elastic couplings are optimized simultaneously to minimize the cost of transport COT. As a result of the optimization, the natural frequency of the robot's swing leg matches with the current step frequency. The robot is then able to walk near its mechanical resonance and has therefore different optimal step lengths and postures to walk with a high efficiency at different walking speeds. The change of robot's posture shifts the optimized equilibrium point of the elastic coupling. In case of a linear torsion spring, the equilibrium point is given by φ_0. Since the characteristic of a CE can be modified by the parameters a and η, its equilibrium point can be adjusted by changing a and η. As depicted in Fig. 9 (top), both types of elastic couplings between the upper body and the thighs (Fig. 1c) result in over 70% reduction of energy consumption, compared to the robot without elastic couplings. As the active deflection of the elastic couplings is generally small (max. 12.31° of the applied CE at a walking speed of 1.4 m/s), the nonlinear CE has a similar performance as the linear torsion spring in terms of improving the energy efficiency. However, the CEs with the parameters in Fig. 9 (bottom) offer a design that can be manufactured and adjusted to the desired walking speed, while changing the preload φ_0 of the linear torsion spring in Fig. 9 (middle) is purely theoretical.

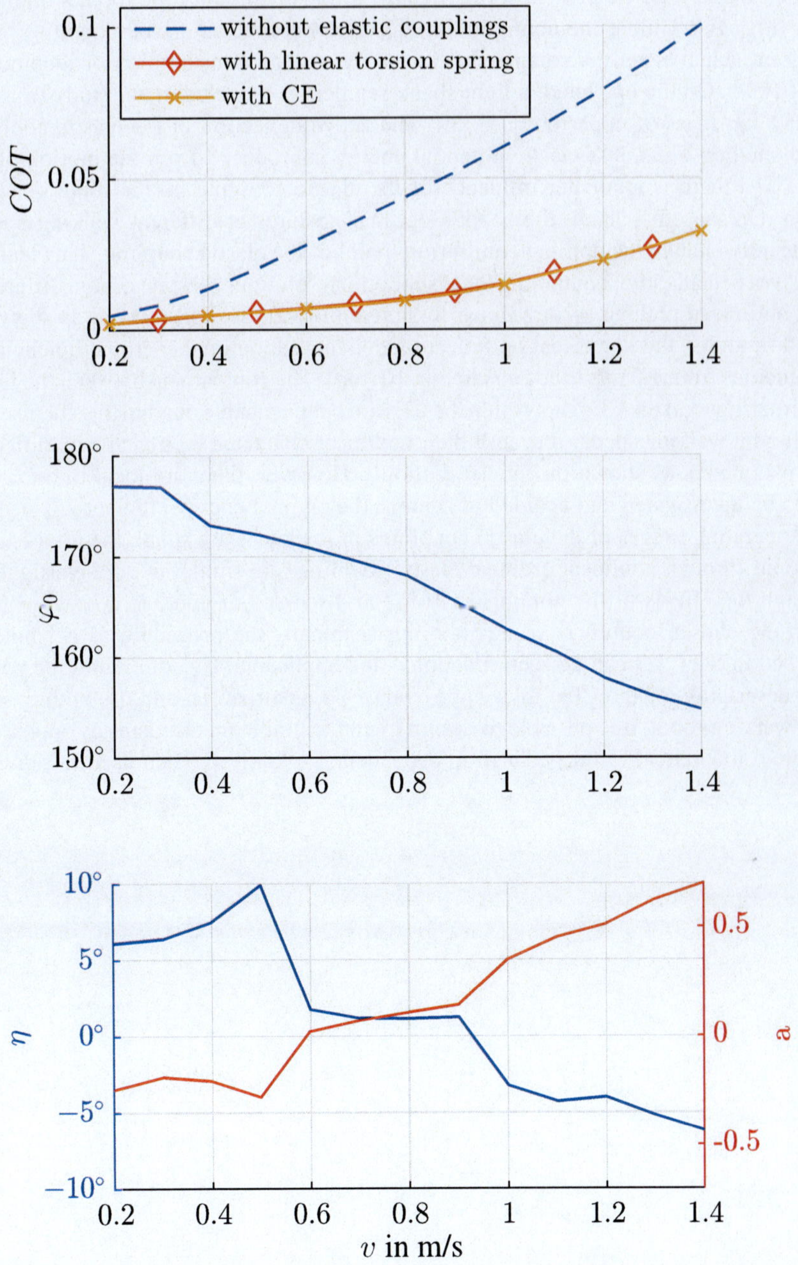

Fig. 9. Top: Cost of transport with and without elastic couplings between robot's upper body and thighs evaluated for walking speeds 0.2 m/s to 1.4 m/s. Middle: Optimized preload φ_0 of a linear torsion spring over different walking speeds. Bottom: Optimized coefficient a (increasing from left to right) and η of a CE over different walking speeds.

In the same way, we present further results of the coupling of the robot's thigh and shank (Fig. 1b). Unlike the configuration described above, an elastic coupling in the knee joint achieves only a small reduction in the energy consumption of locomotion (about 10%). On the one hand, a light shank segment is modeled in our study (mass of shank: 1 kg; mass of upper body: 6 kg). The active deflection of the elastic coupling is also smaller. Thus, less elastic potential energy is produced from the motion of the shanks and there is a smaller influence of the elastic coupling on the robot's natural motion. On the other hand, the robot's optimal postures at different walking speeds significantly change the optimal equilibrium point of the elastic coupling. This leads to excessive specialization containing local suboptimal solutions: despite a large difference of the optimized preload φ_0 of a linear torsion at different walking speeds as shown in Fig. 10 (middle), the numerical optimization delivers suboptimal energy efficiencies at the velocities from 1.0 m/s to 1.4 m/s in Fig. 10 (top). The comparison between the linear torsion springs and the CEs shows that the CEs are better capable of adjusting themselves for different walking speeds, because their nonlinear characteristics can be modified by more parameters as shown in Fig. 10 (bottom). However, there are local minima that have to be investigated and avoided to achieve the optimal energy efficiency. Since the active operating ranges of the elastic couplings in the knees are small, the total energy reductions through nonlinear or linear elastic couplings are similar in our investigation. Although the effects of the nonlinear CE in the discussed example is not very strong due to the chosen couplings of neighboring segments, the procedure as a whole as described in Sect. 3.3 can be seen functional and applicable. By combining the model based description of the CEs, their approximate parametrization and the optimization of the walking robot, it is possible to design manufacturable mechanisms as opposed to the purely theoretical torque-deflection dependencies that have been used in previous works.

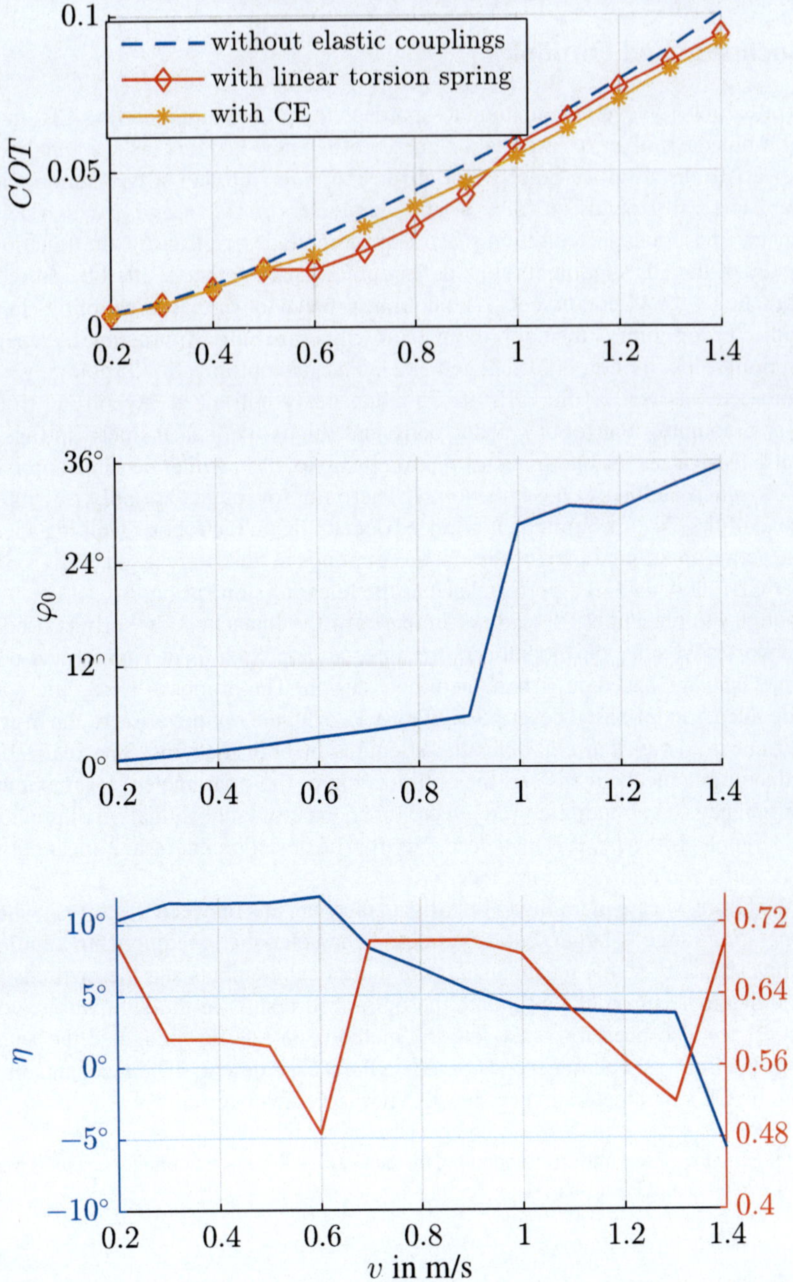

Fig. 10. Top: Cost of transport with or without elastic couplings in robot's knees evaluated for walking speeds 0.2 m/s to 1.4 m/s. Middle: Optimized preload φ_0 of a linear torsion spring over different walking speeds. Bottom: Optimized coefficients a and η (decreases from left to right) of a CE over different walking speeds.

4 Conclusion and Outlook

Our overall aim is to develop a method to approach the problem of inserting CEs in the joints of a bipedal walker to improve its energy efficiency. For this task, an analytical model of a CE based on an Euler-Bernoulli beam is derived and solved numerically in Matlab. The CE depends on four variable parameters ρ_k, κ_0, η and a which define its geometry and boundary conditions. An approximate description for the functional dependency of the CE's moment on its deflection and parameters is fitted based on an ansatz function with 44 coefficients. The nonlinear behavior of elastic couplings in the robot's knee or hip joint is then approximated by this function. Minimizing the energy consumption of locomotion is formulated as a numerical optimization problem, where the robot's gaits as well as the CEs are simultaneously optimized. According to our simulations, coupling the robot's upper body and thighs with ideal linear springs or nonlinear CEs reduces the energy consumption by up to 70%, while the same approach yields 10% when applied to the knee joints. These improvements are only possible if the preload of the elastic couplings is adjusted depending on the robot's walking speed. Since the active operating range of the elastic couplings in this study is generally small, nonlinear CEs show a similar performance as the linear torsion springs.

Although the presented CEs do not outperform the linear torsion springs used in previous works, we have developed and presented an approach for designing and optimizing mechanisms based on a manufacturable design. The proposed procedure combines a detailed, model-based description of the CEs with the optimization of the overall robot system in a way that combines the strengths of both methods. The realization of manufacturable mechanisms and the optimization of their parameters for maximum energy efficiency is an important prerequisite for an experimental validation of previous works [7, 8] in a prototype. It is also necessary for the development of nonlinear elastic couplings with even higher performance.

For the future, we are planning to investigate mechanisms between the thighs, where active operating range is larger, but a symmetric characteristic is required. In addition, we will investigate bi-stable mechanisms with compliant elements and the methods for their modifications. In order to validate the optimized results in the simulations with a prototype, we also need to investigate the methods on how to recognize the actual operating state such as walking speeds based on the measurements. The mechanisms of adjusting the CE's parameters in practice also has to be investigated.

Acknowledgement. The authors are supported by the German Research Foundation (DFG) grant number FI 1761/4-1 and ZE 714/16-1.

References

1. Seok, S., Wang, A., Chuah, M.Y., Hyun, D.J., Lee, J., Otten, D.M., Lang, J.H., Kim, S.: Design principles for energy-efficient legged locomotion and implementation on the MIT Cheetah Robot. IEEE/ASME Trans. Mechatron. (2015). https://doi.org/10.1109/tmech.2014.2339013
2. Nasiri, R., Zare, A., Mohseni, O., Yazdanpanah, M.J., Ahmadabadi, M.N.: Concurrent design of controller and passive elements for robots with impulsive actuation systems. Control Eng. Pract. (2019). https://doi.org/10.1016/j.conengprac.2019.03.014

3. Sharbafi, M.A., Yazdanpanah, M.J., Ahmadabadi, M.N., Seyfarth, A.: Parallel compliance design for increasing robustness and efficiency in legged locomotion—proof of concept. IEEE/ASME Trans. Mechatron. (2019). https://doi.org/10.1109/TMECH.2019.2917416
4. Hobart, C.G., Mazumdar, A., Spencer, S.J., Quigley, M., Smith, J.P., Bertrand, S., Pratt, J., Kuehl, M., Buerger, S.P.: Achieving versatile energy efficiency with the WANDERER biped robot. IEEE Trans. Robot. (2020). https://doi.org/10.1109/tro.2020.2969017
5. Hubicki, C., Abate, A., Clary, P., Rezazadeh, S., Jones, M., Peekema, A., van Why, J., Domres, R., Wu, A., Martin, W., Geyer, H., Hurst, J.: Walking and running with passive compliance: lessons from engineering: a live demonstration of the ATRIAS biped. IEEE Robot. Automat. Mag. (2018). https://doi.org/10.1109/MRA.2017.2783922
6. Iida, F., Rummel, J., Seyfarth, A.: Bipedal walking and running with compliant legs. In: Proceedings. 2007 IEEE International Conference on Robotics and Automation. Proceedings 2007 IEEE International Conference on Robotics and Automation, 2007/04. IEEE Press Books, New York (2007). https://doi.org/10.1109/robot.2007.364088
7. Bauer, F., Römer, U., Fidlin, A., Seemann, W.: Optimization of energy efficiency of walking bipedal robots by use of elastic couplings in the form of mechanical springs. Nonlinear Dyn. **83**(3), 1275–1301 (2015). https://doi.org/10.1007/s11071-015-2402-9
8. Bauer, F., Römer, U., Fidlin, A., Seemann, W.: Optimal elastic coupling in form of one mechanical spring to improve energy efficiency of walking bipedal robots. Multibody Sys.Dyn. **38**(3), 227–262 (2016). https://doi.org/10.1007/s11044-016-9509-8
9. Ding, J., Zhou, C., Xiao, X.: Energy-efficient bipedal gait pattern generation via CoM acceleration optimization. In: 2018 IEEE RAS 18th International Conference on Humanoid Robots (Humanoids). 2018 IEEE-RAS 18th International Conference on Humanoid Robots (Humanoids), pp. 238–244 (2018). https://doi.org/10.1109/HUMANOIDS.2018.8625042
10. Chen, T., Schmiedeler, J.P., Goodwine, B.: Robustness and efficiency insights from a mechanical coupling metric for ankle-actuated biped robots. Auton. Rob. **44**(2), 281–295 (2019). https://doi.org/10.1007/s10514-019-09893-w
11. Römer, U., Fidlin, A., Seemann, W.: Investigation of optimal bipedal walking gaits subject to different energy-based objective functions. Pamm **15**, 69–70 (2015)
12. Mazumdar, A., Spencer, S.J., Hobart, C., Salton, J., Quigley, M., Wu, T., Bertrand, S., Pratt, J., Buerger, S.P.: Parallel elastic elements improve energy efficiency on the steppr bipedal walking robot. IEEE/ASME Trans. Mechatron. **22**, 898–908 (2017)
13. Howell, L.L.: Compliant Mechanisms. Wiley, New York (2001)
14. Zentner, L.: Nachgiebige Mechanismen. De Gruyter, München (2014)
15. Howell, L.L., Magleby, S.P., Olsen, B.M. (eds.): Handbook of Compliant Mechanisms. Wiley, Chichester (2013)
16. Lobontiu, N.: Compliant Mechanisms. Design of Flexure Hinges. CRC Press, Boca Raton (2002)
17. Zentner, L., Linß, S.: Compliant systems. Mechanics of Flexible Mechanisms, Actuators and Sensors. De Gruyter Oldenbourg, Berlin, Boston (2019)
18. Henning, S., Linß, S., Vollrath, T., Zentner, L.: Elasto-kinematic modeling of planar flexure hinge-based compliant mechanisms incorporating branched links. In: Uhl, T. (ed.) Advances in Mechanism and Machine Science. Mechanisms and Machine Science, vol. 44, pp. 1599–1608. Springer, Cham (2019)
19. Henning, S., Linß, S., Zentner, L.: detasFLEX – a computational design tool for the analysis of various notch flexure hinges based on non-linear modeling. Mech. Sci. (2018). https://doi.org/10.5194/ms-9-389-2018

Check for updates

Modeling, Design and Prototyping of a Pantograph-Based Compliant Mechanism

Dušan Stojiljković[1(✉)], Lena Zentner[2], Nenad T. Pavlović[1], Sebastian Linß[2], and René Uhlig[2]

[1] Faculty of Mechanical Engineering, University of Niš, Niš, Serbia
dusan.stojiljkovic@masfak.ni.ac.rs

[2] Compliant Systems Group, Technische Universität Ilmenau, Ilmenau, Germany

Abstract. The main task of the compliant mechanism synthesis is to generate a pre-defined motion path as accurately as possible. A general approach to the compliant mechanism synthesis is to develop a compliant mechanism based on the rigid-body model by replacing conventional joints with compliant joints, i.e. flexure hinges. Using the example of a mechanism producing a scissors-like motion, in this paper a more specific and iterative synthesis process is implemented for the design of a compliant path-generating mechanism. Based on two symmetric pantograph mechanisms, a kinematic analysis of the multi-link rigid-body model is performed. The final dimensions and link lengths of the rigid-body model are used to implement a compliant mechanism with different flexure hinges. Therefore, several designs are iteratively investigated by means of FEM simulations in order to improve the path accuracy and the opening angle of the scissors-like motion.

Keywords: Compliant mechanism · Mechanism design · Pantograph mechanism · Scissors-like motion · 3D printing

1 Introduction

Compliant joints gain all of their motion and transmitting force only from their elasticity, and because of this elastic deformation, the motion range of these flexible segments is restricted [1]. The movement of these segments is very limited and leads to small deviations and displacements of compliant mechanisms. However, there are many advantages of using compliant joints over their rigid-body counterparts used in classical mechanisms; they have less wear, weight, noise and backlash. Duo to this and many other advantages, compliant mechanisms are used in semiconductor and micromechanics technologies [2]. Compliant mechanisms are functionally similar to classical (rigid-body) mechanisms, but their mobility is based on the deformation of the flexible members and not on the mobility of the joints themselves [3]. However, since compliant mechanisms are relatively new compared to traditional mechanisms, they are difficult to design and it

L. Zentner and S. Strehle (Eds.): MAMM 2020, MMS 96, pp. 76–88, 2021.
https://doi.org/10.1007/978-3-030-61652-6_7

is not easy to find examples and resources that will lead to a simple replacement of traditional mechanisms by compliant ones. This has a particular effect on the mechanisms that provide a large motion range of their segments.

There are many papers that offer information on large motion compliant mechanisms. Paper [4] introduces metrics for large-displacement linear-motion compliant mechanisms. Paper [5] presents the topology synthesis of large displacement compliant mechanisms with specific output motion paths. There are also many articles, texts and student theses that investigate large-range compliant mechanisms [6–8].

However, not enough literature is dedicated to the consideration of the pantograph apparatus as a basis for developing a compliant mechanism. Paper [9] describes a statically balanced concept and demonstrates its optimization, testing, and implementation for a haptic pantograph mechanism. In papers [10, 11] it can be seen how compliant pantograph mechanisms can be used for linear displacement applications. Paper [10] exhibits a large motion displacement analyzed with an FEM simulation. For the model shown in paper [12], the pantograph apparatus is not included but a similar double parallel mechanism is shown and used to create a specific desired motion.

This paper deals with a similar type of motion presented through a mechanism with two symmetrical pantograph mechanisms. Hence, a compliant mechanism was designed using an iterative synthesis approach so as to achieve large scissors-like motion. This paper aims to employ analysis of thus given mechanisms providing large output motion, as well as the stationary rotation of the coupler point of two neighboring rocker links.

2 Modeling of the Rigid-Body Mechanism

The pilot mechanism examined in this paper is obtained using the basic function of the pantograph apparatus. A mimic of rotation around the support translates into even greater rotation at the outlet of two symmetrical pantograph mechanisms (Fig. 1a). Mapping this mechanism and merging the last segments in the center of the common circle results in a stationary rotation axis of two neighboring rocker links that are not connected to the frame (Fig. 1b).

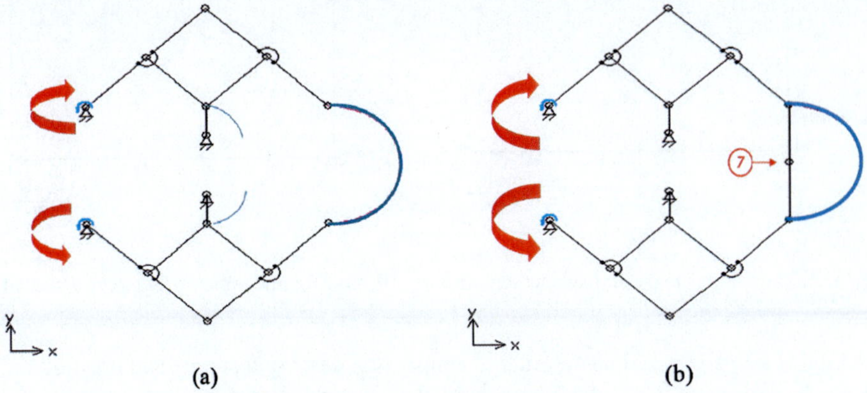

Fig. 1. (a) Two pantograph mechanisms, (b) Joint 7 of two pantograph mechanisms

Figures 1a and 1b show that the mechanism thus obtained resembles the scissors-like motion, which will affect the appearance of the final model at the last stage of production.

Adding segments to the given mechanism, by which translational movement is obtained, represents an input to track the scissors-like motion (joint 10 in Fig. 2). By moving joint 10, the angle at the output can be increased, but its position must be carefully chosen so that no other direction of rotation of the segments occurs (the leading segments 9 and 22 would move in the directions opposite to the desired ones, causing the additional closing of segments 7 and 14 at the exit) [13].

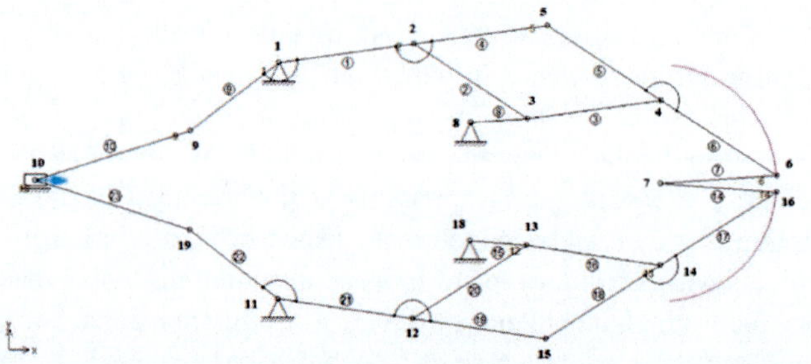

Fig. 2. Rigid-body mechanism in its initial position

After performing a kinematic analysis in SAM 6.1, a diagram of the input translational movement is obtained (Fig. 3a). The diagram shows that the input movement, the movement of point 10, is 5.22 mm, which gives an output movement of the segments that form an angle of 90° (Fig. 3b).

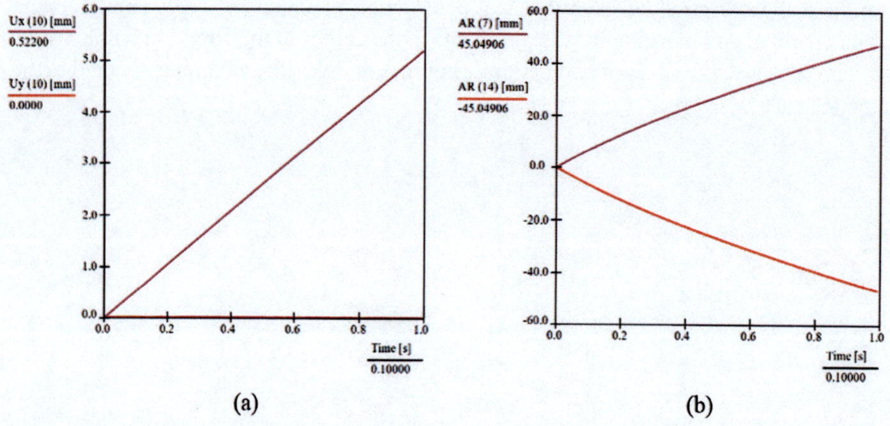

Fig. 3. (a) Graph of the input displacement of point 10, (b) Moving angle of segments 7 and 14

All this analysis was done in order to confirm the desired accuracy and operation of the mechanism itself. Consideration should also be given to the function of the pantograph, which allows the changing of the pantograph joint position to change the scale in which the original drawing is conveyed when drawing a copy thereof. In this case, it means that if the position of the pantograph joints is such that the ratio is 1:3, the circular arc described around support 8, whose new diameter is 20 mm, will be mapped into a circular arc with a diameter of 60 mm centered at point 7 (Fig. 4).

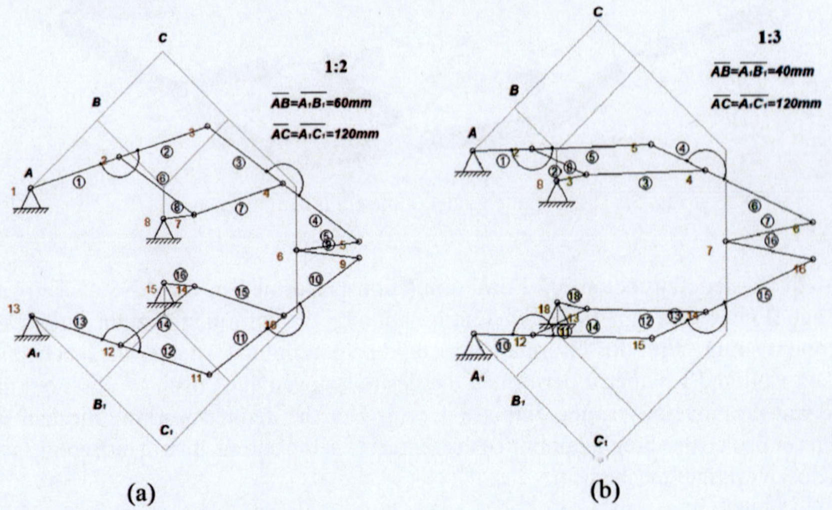

Fig. 4. The initial (black color) and final position (gray color) of the mechanism with ratio: (a) 1:2, (b) 1:3

Figure 4 shows that the ratio given by the shape of the pantograph affects the length of the rigid segment at the output. This is a positive factor, but it brings with it a disadvantage, which is that by increasing the length of the solid segment, the angle that that segment describes is reduced for the same input value. The attempts were made to reduce the ratio to 1:2.5, but the same problem arose. Knowing this, it can be concluded that the best version of the pantograph would be with a 1:2 ratio for this case.

3 Design of the Compliant Mechanism

For further analysis, an initial working version of the mechanism was modeled, forming the basis for the investigation of the mechanism operation as well as its greater operational precision (Fig. 5). The dimensions and the height of the model and the shape of the compliant joints were taken arbitrarily. As such, they provide a good basis for further work and analysis, as well as a way to find the best solution for the parameters of this model and the underlying mechanism [14].

All compliant joints were taken as rectangular notches or leaf-type notch hinges. Subsequently, all joints of this shape were used because of its simple modeling ability as well as small dimensions at large deflections, which was necessary for this work.

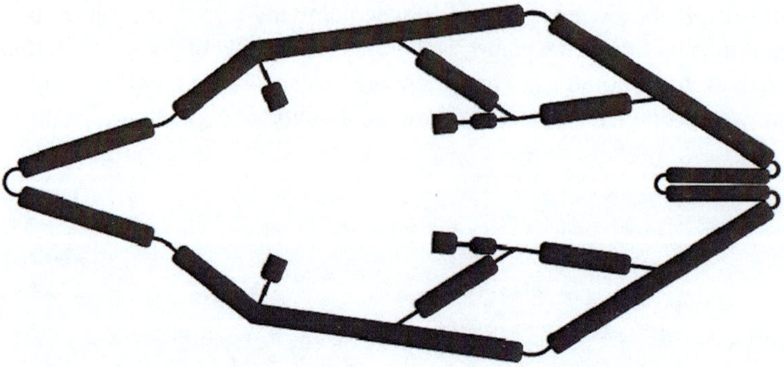

Fig. 5. The working model of the compliant mechanism

A quasi-static structural 3D FEM simulation performed in the ANSYS software package (Figs. 6 and Fig. 7) provided information on the deformation at the output and maximum strain. With this information it could be determined whether the mechanism worked well and whether it performed the desired movement. It can be observed that there was no movement at the output, i.e. joint 7 of the defined working mechanism, which confirmed the circular motion of the segments at the output, thus in turn confirming the chosen original mechanism.

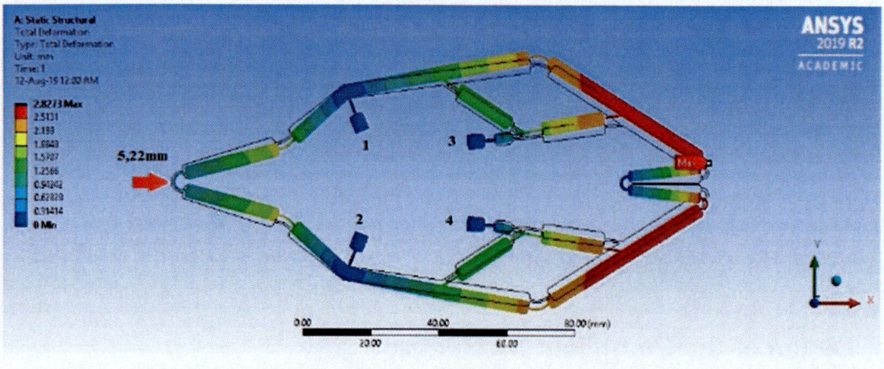

Fig. 6. FEM results of the deformation with boundary conditions of the model

The maximum strain was only 2% at two symmetrical compliant joints (Joints 9 and 19 of the defined rigid-body mechanism), as can be seen in Fig. 7. This is a good basis for checking the function of compliant joints and also getting the best solution for their shape and width. All the compliant joints below are rectangular joints with the width of 0.7 mm in the Z direction.

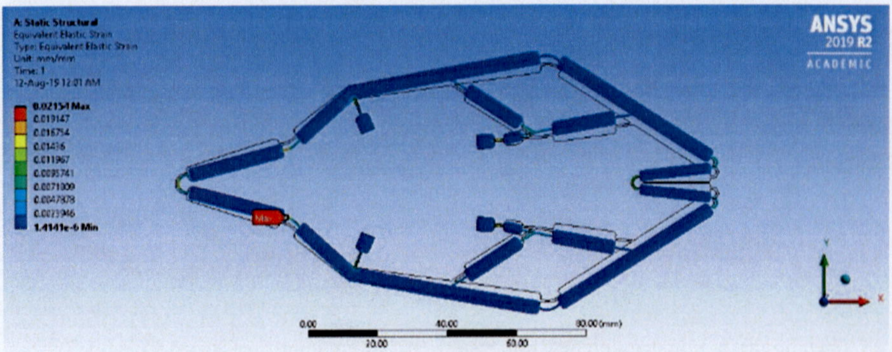

Fig. 7. FEM results of the equivalent von-Mises strain of the model

3.1 The 1st Iteration Model

The first step before modeling was to calculate the length of the 0.7 mm thick compliant joints. The minimum joint length values were obtained by using the detasFLEX program, developed in MATLAB [15]. Table 1 shows the values of all joints obtained by calculation.

Table 1. Calculation of compliant joints

Joint	Angle [°]	Length [mm]
1,11	20.94	7
2,12	25.28	8.5
3,13	49.39	16.5
3,13	25.28	8.5
3,13	24.12	8.5
4,14	25.28	8.5
5,15	25.28	8.5
6,16	49.40	16.5
7	90.10	30
8,18	45.05	15.5
9,19	30.17	10.5
10	18.47	6.5

Figure 8 shows a model with new dimensions of the rigid segments and compliant joints, as well as an analysis showing the final position occupied by this mechanism.

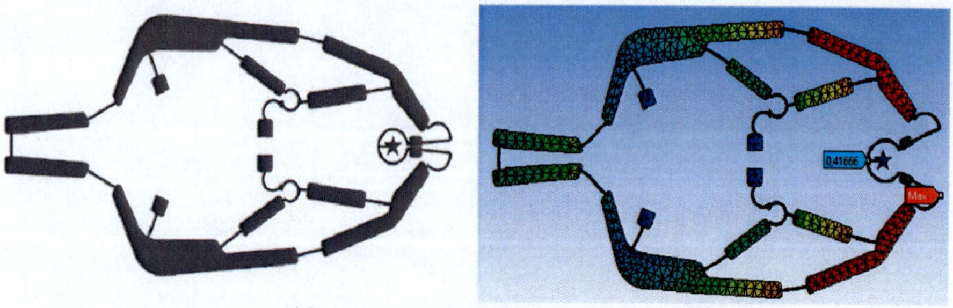

(a) (b)

Fig. 8. (a) The model with new dimension, (b) Deformation analysis of the model

By observing this model, it was concluded that a change in the position of joint 10 (input) could have an effect on improving the operation of the mechanism: a smaller input displacement and a higher angle at the output (Fig. 9). Also, for easier analysis, the shape of the compliant joints at the outlet was changed (Fig. 10).

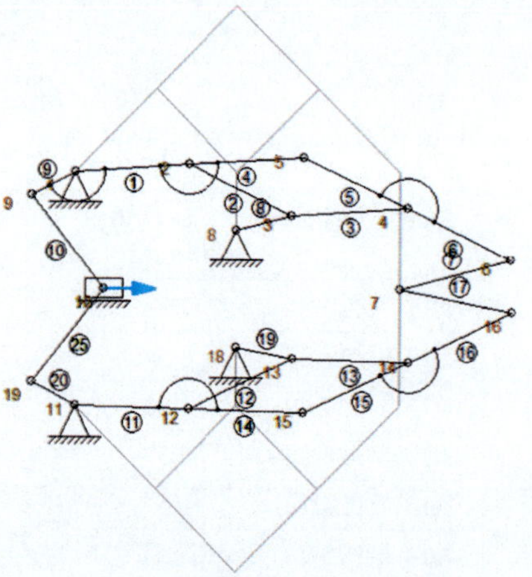

Fig. 9. The 1st iteration mechanism

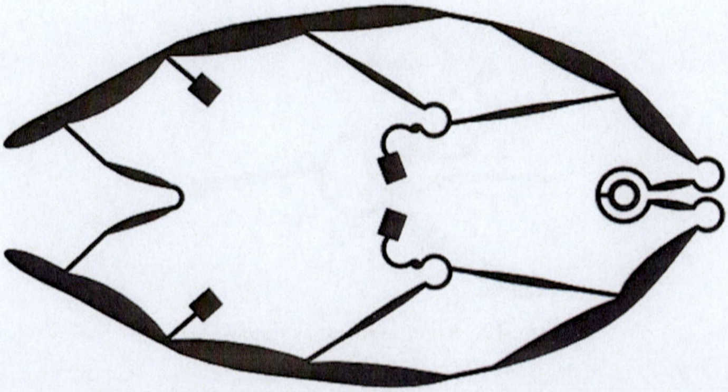

Fig. 10. The 1st iteration model with new compliant joints and rigid segments

3.2 The 2nd Iteration Model

As the end result of the 1st iteration model was not satisfactory, the 2nd iteration model was modeled similarly to the 1st model but with a new type of translational input. The compliant translational input was modeled from the role model presented in paper [16]. Figure 11 shows the 2nd iteration model prepared for 3D printing.

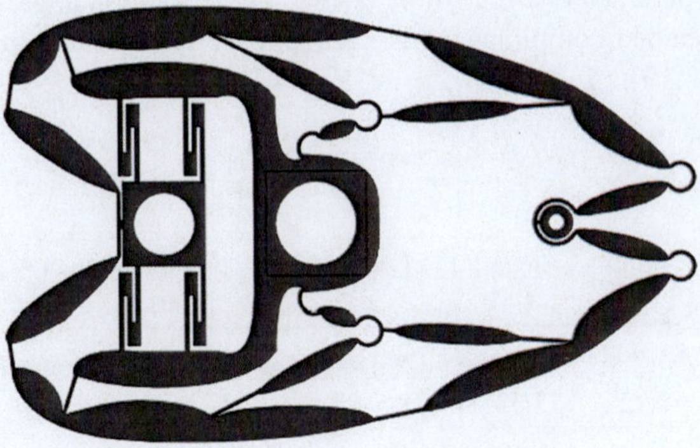

Fig. 11. The 2nd iteration model

This model did not deliver good performance due to the translational input that was not mobile enough and because of the small output angle. The solution was to find a new type of translational input.

3.3 The 3rd Iteration Model

As the model dimensions were not suitable for modeling rectangular compliant joints without arched curves, this was the first parameter to be changed. The length of the rigid segment describing the circular arc around support 8 was also changed from 12 mm to 20 mm in order to model a larger rigid segment at that position compared to the previous case (Fig. 12).

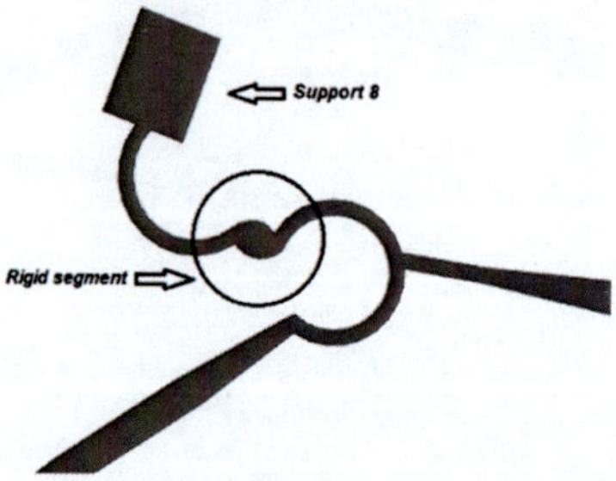

Fig. 12. Rigid segment at support 8

Because of this change, and also the aesthetics of the model itself, the rigid segments were changed from 28 mm to 40 mm. As the two previous models gave poor results, the following changes needed to be implemented in the new modeling:

- all compliant joints were modeled in a rectangular shape;
- the length of compliant joint 8 was changed from 12 mm to 20 mm;
- the length of rigid segments was changed from 28 mm to 40 mm;
- the output angle was changed from 90° to 60°.

All the above parameters affected the change in the rotation angles of the joints so that the newly obtained compliant joints needed to be recalculated in the detasFLEX program (Table 2) [15].

Table 2. Calculation of compliant joints for the 3rd model

Joints	Angle [°]	Length [mm]	Real Length [mm]
1, 11	19.14	7	7
2,12	25.53	8.5	8.5
3,13	36.68	12	6 × 6 × 2.5
4,14	25.53	8.5	8.5
5,15	25.53	8.5	8.5
6,16	36.68	12	12
7	60.58	20	5.95 × 17.75
8,18	30.29	10	10
9,19	10.80	4	4
10	16.69	6	12

The new model of the mechanism caused changes in the appearance of the compliant joints as mentioned earlier. Likewise, the rigid segments were enlarged to reduce the flexibility of the whole model in undesired directions (Fig. 13).

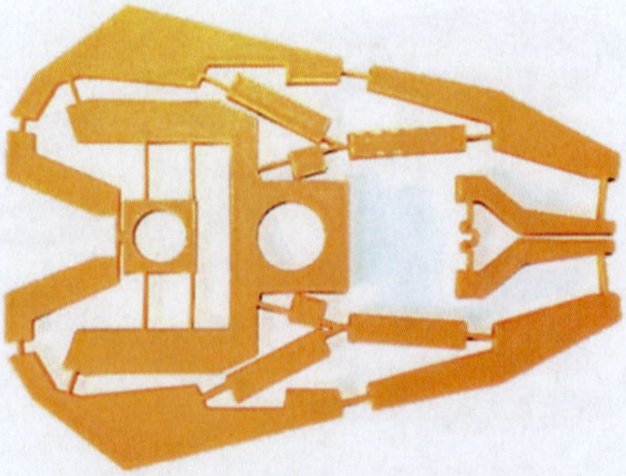

Fig. 13. The 3rd iteration model

The compliant translational input needed to be changed as well because the existing one did not give enough stroke to achieve the desired input. It also needed to be modeled with a more suitable shape of joint 7.

3.4 The 4th Iteration Model

Joining the bends on both sides, of the 2nd iteration model, into one enlarged bend produced a new compliant translational input. The analysis showed that this design was very suitable for this application. For a very small input displacement force, high mobility in the horizontal direction was obtained (Fig. 14). Strain values were very small, so this compliant translational input proved to be suitable.

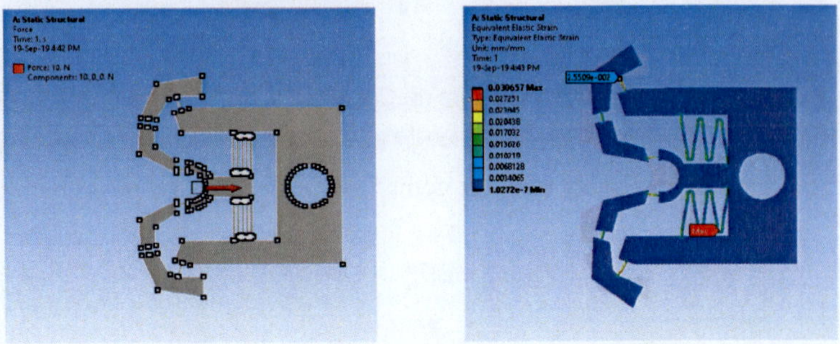

(a) (b)

Fig. 14. FEM analysis of the compliant translational input: (a) Input force, (b) Strain

The analysis of the previous models concludes that the latest version, the 4th iteration model, gives the best results. Only the shape of joint 7 was retained to provide a clearer picture of the rotating joint (Fig. 15).

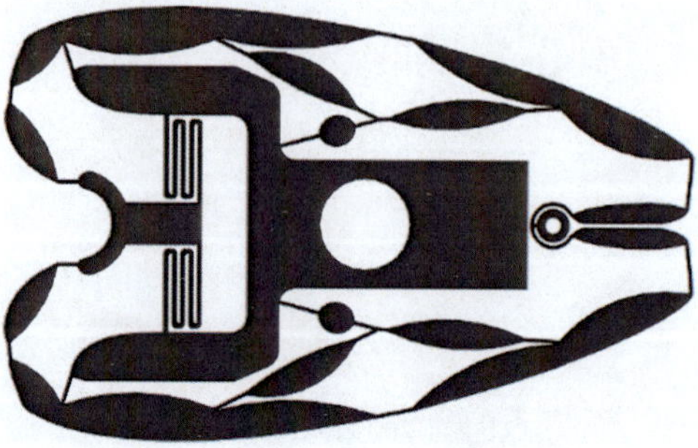

Fig. 15. The 4th iteration model

4 Prototyping of the Compliant Mechanism

The modeling of the final compliant mechanism can be seen in Fig. 16. In addition to all changes and modifications already mentioned, certain details were added to produce a representation of a fish with fins.

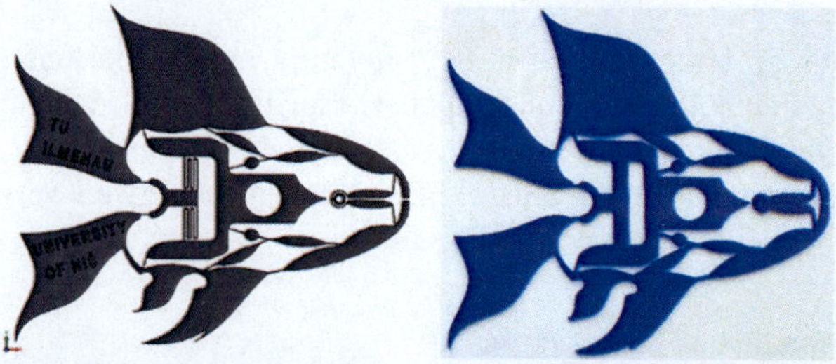

(a) (b)

Fig. 16. Final compliant mechanism (a) CAD model, (b) 3D printed model

The results of an FEM analysis of the compliant mechanism can be seen in Fig. 17.

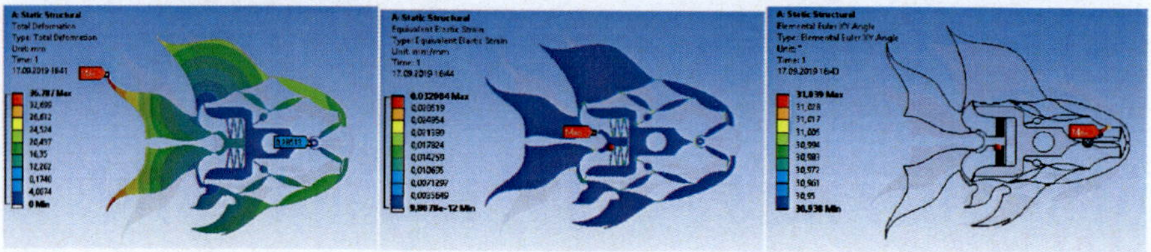

(a) (b) (c)

Fig. 17. FEM analysis of the final compliant mechanism: (a) Total deformation, (b) Strain, (c) Value of the output angle

As can be seen from the analysis, this model gives the best values of input displacement, output displacement and minimal translation of joint 7 [17].

5 Conclusion

This paper emphasized the development of a compliant mechanism that produces a scissors-like output motion. Therefore, an iterative synthesis process was applied based on the FEM analysis and critical verification of each model in order to achieve the desired goal of a quasi-constant rotational joint which was not fixed to the frame. Several designs were iteratively investigated to improve the path accuracy of the used pantograph mechanism.

Throughout the development, the process of modeling, design and fabrication of the compliant mechanism is reported. The final fish-looking model is a suitable solution due to its unique functionality and the desired specific movement of the mechanism output segments, represented as the fish mouth: a small input displacement and a large output opening angle of the rotating mouth segments while the undesired displacement of the connecting hinge can be neglected.

Acknowledgments. This research was conducted within the project "Synthesis, realization and control of different bio-inspired spatial compliant systems with structurally integrated highly elastic sensors and actuators" (bilateral scientific and research cooperation program Serbia - Germany 2018–2019). This research was financially supported by the Ministry of Education, Science and Technological Development of the Republic of Serbia.

References

1. Zentner, L., Linβ, S.: Compliant systems - Mechanics of Elastically Deformable Mechanisms, Actuators and Sensors, De Gruyter, Oldenbourg, München, April,2019
2. Pavlović, N.D., Pavlović, N.T.: Gipki mehanizmi [Compliant mechanisms], Mašinski fakultet Univerziteta u Nišu (2013). ISBN 978-86-6055-036-3
3. Howell, L.L., Magleby, S.P., Olsen, B.M.: Handbook of Compliant Mechanisms. John Wiley & Sons Inc., New York (2012)
4. Mackay, A.B., Smith, D.G., Magleby, S.P., Jensen, B.D., Howell, L.L.: Metrics for evaluation and design of large-displacement linear-motion compliant mechanisms. J. Mech. Des. **134**(1), 011008 (2012). https://doi.org/10.1115/1.4004191
5. Dirksen, F., Berg, T., Lammering, R., Zohdi, T.I.: Topology synthesis of large-displacement compliant mechanisms with specific output motion paths. In: Proceedings in Applied Mathematics and Mechanics, vol. 12, no. 1, pp. 801–804, 3 December 2012. https://doi.org/10.1002/pamm.201210384
6. Mackay, A.B.: Large-Displacement Linear-Motion Compliant Mechanisms, Master's Thesis, Brigham Young University - Provo (2007)
7. Pei, X., Yu, J., Zong, G., Bi, S.: An effective pseudo-rigid-body method for beam-based compliant mechanisms. Precis. Eng. **34**(3), 634–639 (2010)
8. Xu, Q.: Design and Implementation of Large-Range Compliant Micropositioning Systems. John Wiley & Sons Inc., New York (2016)
9. Merriam, E. G., Colton, M., Magleby, S. P., Howell, L. L.: The design of a fully compliant statically balanced mechanism. In: Proceedings of the ASME 2013 International Design Engineering Technical Conferences and Computers and Information in Engineering Conference. Volume 6A: 37th Mechanisms and Robotics Conference, Portland, Oregon, USA, 4–7 August 2013. https://doi.org/10.1115/DETC2013-13142

10. Patil, V., Anerao, P.R., Chinchanika, S.S.: Finite element analysis of compliant pantograph mechanism. Int. Eng. Res. J. pp. 65–68 (2016)
11. Deshmukh, B., Pardeshi, S., Mishra, P.K.: Conceptual design of a compliant pantograph. Int. J. Emer. Technol. Adv. Eng. **2**(8), 270–275 (2012)
12. https://www.compliantmechanisms.byu.edu/wyrd-mechanisms
13. Pavlović, N.D., Milošević, M.: Polužni mehanizmi [Linkage mechanisms], Mašinski fakultet Univerziteta u Nišu (2012). ISBN 978-86-6055-029-5
14. Pavlović, N.T., Pavlović, N.D.: Mobility of the compliant joints and compliant mechanisms. Theor. Appl. Mech. **32**(4), 341–357 (2005)
15. Henning, S., Linß, S., Zentner, L.: detasFLEX – a computational design tool for the analysis of various notch flexure hinges based on non-linear modeling. Mech. Sci. **9**, 389–404 (2018). https://doi.org/10.5194/ms-9-389-2018
16. Zhang, X., Xu, Q.: Design and analysis of a 2-DOF compliant gripper with constant-force flexure mechanism. J. Micro-Bio Robotics **15**(1), 31–42 (2019)
17. Stojiljković, D.S.: Modeliranje i izrada gipkog mehanizma skeleta ribe [Modeling and fabrication of a compliant fish skeleton mechanism]. Mašinski fakultet Univerziteta u Nišu, Završni (master) rad (2019)

Wafer Bonding in MEMS Technologies

Roy Knechtel[1(✉)] and Uwe Schwarz[2]

[1] Schmalkalden University of Applied Sciences Autonomous Intelligent Sensors, Chair of the Carl-Zeiss-Foundation, Schmalkalden 98574, Germany
r.knechtel@hs-sm.de
[2] X-FAB MEMS Foundry GmbH, Erfurt 99097, Germany

Abstract. Wafer bonding is the stacking and joining of semiconductor substrates and is an essential process step in the development and production of MEMS (micro-electrical-mechanical systems) [1]. It provides the possibility to realize real three-dimensional structures, in this technology field where 2D-strutures with fixed thickness are still dominating. Actually nano- and micro mechanical and electrical elements are required and possible to be bonded at wafer level, this finally allows system integration across functional and geometrical domains.

Keywords: Wafer bonding · MEMS · Engineered substrates · Anodic bonding · Direct bonding · Glass frit bonding

1 Introduction

The main purpose of wafer bonding is the realization of three-dimensional structures in the MEMS and related wafer process. This has two main aspects so far (see also Fig. 1): Realization of special substrates with buried layers such as oxide and/or implanted layers in SOI wafers, or buried structures such as sealed cavities in silicon. These special substrates provide advanced properties for device function and enable new features to be possible. For example, dielectrically insulated silicon areas can be realized in SOI wafers, which are important for high voltage applications as well as for MEMS devices. On the other hand, these types of special substrate give many technological advantages, and even enable new and efficient technologies to be possible. Examples include absolute pressure sensors based on substrates with sealed cavities or an etch stop at the buried oxide of SOI wafers. Capping of MEMS [1, 2, 4] and micro fluidic devices at the end of the wafer process is the other important wafer bonding application. MEMS structures, particularly those of surface micromechanical devices must be protected at the wafer level to prevent destruction by mechanical or environmental influences. Hermetic sealing is also often required as an extra effective protection during dicing and assembly processing, as well as for device lifetime and application. This type of wafer bonding does not only provide protection, but also completes the functionality of the chips. Microfluidic channels must be sealed by caps in order to carry fluids, absolute pressure sensors and resonant structures, such as gyroscopes, require a vacuum in their cavity, while in other sensors, the cap is provided as an electrical shielding or counter electrode. Finally,

L. Zentner and S. Strehle (Eds.): MAMM 2020, MMS 96, pp. 89–102, 2021.
https://doi.org/10.1007/978-3-030-61652-6_8

the cap wafer can contain electrical structures as full application specific integrated circuits (ASICs) for sensor signal conditioning - this allows full sensor system integration on chip level by wafer bonding. Wafer boding processes are actually needed to realize modern and high performing microsensors like pressure sensors, acceleration sensors, gyroscopes, gas and gas flow sensors, as well as X-Ray detectors and it drives the integration of sensing elements and electronics like for the sensors just mentioned but also for MEMS Microphones or high performing microfluidics. Also, for micro actuators like micropumps, MEMS-speakers or cell sorting devices wafer bonding is essentially important. Processes and materials from the field of wafer bonding are also used in the classical field of precisions engineering. For examples the glass frit material described in this paper is used to mount silicon based strain gauges to deforming elements made from special steel to the improve mechanical coupling.

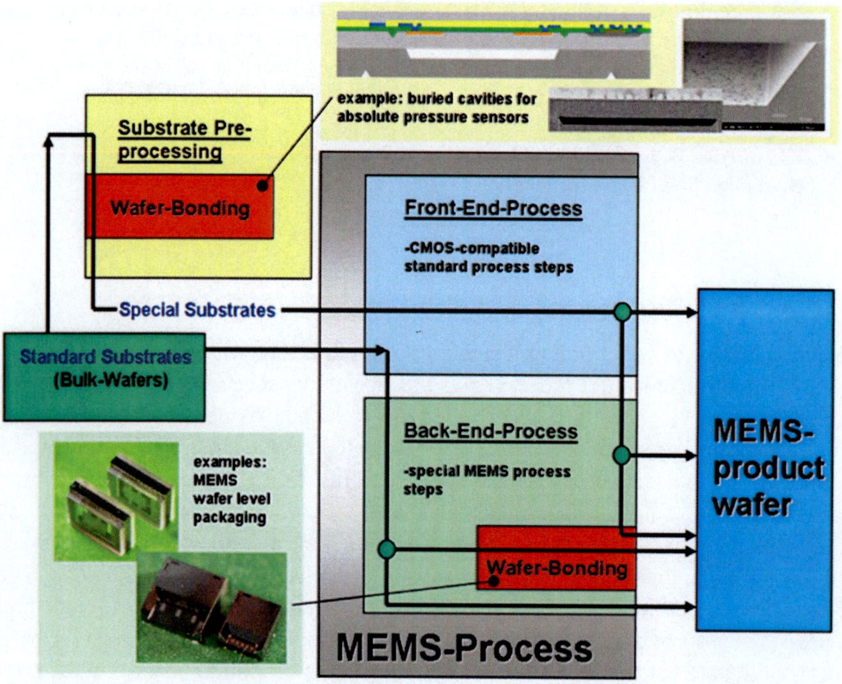

Fig. 1. Usage of wafer bonding in MEMS processing for both substrate pre-processing and wafer level capping

Beside the engineered substrates and wafer level capping and integration aspects the temporary wafer bonding [3] especially for thin wafer handling and processing is getting important for the MEMS processing. To realize the bonding of plain or structured MEMS wafers different technological concepts can be applied, the most important of them are introduced in this paper and are discussed regarding process integration aspects.

2 Engineered Substrates

2.1 High Temperature Semiconductor Wafer Bonding

Engineered substrates are enabler of new and complex technologies. The best-known examples are Silicon on Insulator (SOI) Wafers for modern MOS technologies like for processors and memories. The construction of a thin silicon device layer on top of a buried oxide placed on a handle wafer, provides much benefits like improved temperature and high voltage behaviour and the trench insulation of the electronic circuitries. In MEMS technologies such pre-processed engineered substrates contain mainly additional cavities, which can be advantageously be used as reference pressure cavities for. Absolute pressure sensors or complex mechanical structures can be structured out of thin membranes covering the cavities.

In order to produce special engineered substrates, the technology of choice is high temperature, direct (fusion) wafer bonding. On the one hand, in subsequent processing, the substrates will be heated up to temperatures of around 1000 °C anyway, during thermal oxidation or implantation annealing steps. On the other hand, only minimal processing is applied to the wafers before bonding, so that their surface roughness is not significantly increased and the polished surface remains direct-bondable. Aggressive wet chemical activation of the wafers is possible because, in order to be compatible with high temperature processing, they do not contain metal features. A typical high temperature wafer bonding process consist of three steps:

- Cleaning and activation of the bonding wafers using wet chemicals such as RCA cleaning or other acid combinations, to remove particles which would locally prevent bonding (bond defects) and add hydroxyl groups to the surface (surface hydrophilisation) as conditioning for the pre-bonding.
- Pre-bonding the wafers at room temperature by mechanical pressure, to achieve wafer to wafer contact in the atomic range, so that the hydroxyl groups interact and form an initial, hermetically tight but still weak bond. If the pre-bond is performed in a bonding chamber, vacuum or gasses can be sealed inside the cavities.
- At high temperature annealing of about 1000 °C, the bond interface becomes strong, of the order of crystalline silicon. By outgassing of water from the bond interface, a stable oxygen bond is formed, which is manifested as a thin oxide layer of a few nanometer thickness in cross sections.

Figure 2 shows a typical pre-bonding of structured wafers as a moving bond front, while Fig. 3 illustrates the chemical effects occurring during the three bonding steps.

Through high temperature direct bonding, several surface layers, such as hydrophilic silicon with or without deep etched cavities, thermal oxide, implanted layers, silicides (e.g. CoSi2 [5]) and others can be bonded. This provides a very wide range of possibilities for substrate engineering for special applications. The bond process as described above is very reproducible and safe if the surface roughness is sufficiently low (Ra $<$ 2 nm), and this requirement is very critical in practice. Prime polished wafers normally have a low enough roughness, but this will be increased during processing, which may result in the wafers becoming unbondable. If chemical mechanical polishing (CMP) cannot

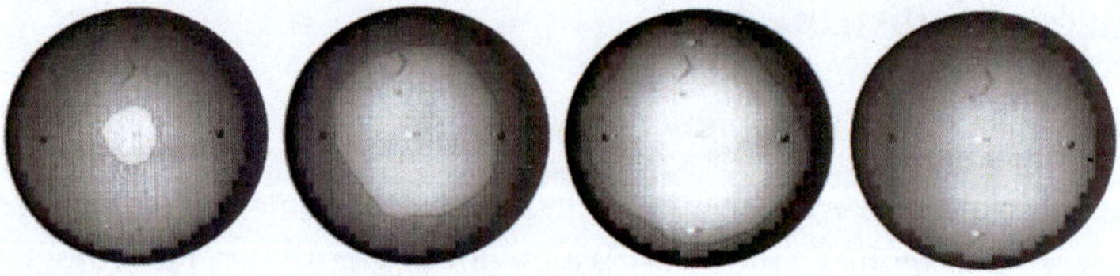

Fig. 2. Moving bond front during pre-bonding of structured wafers

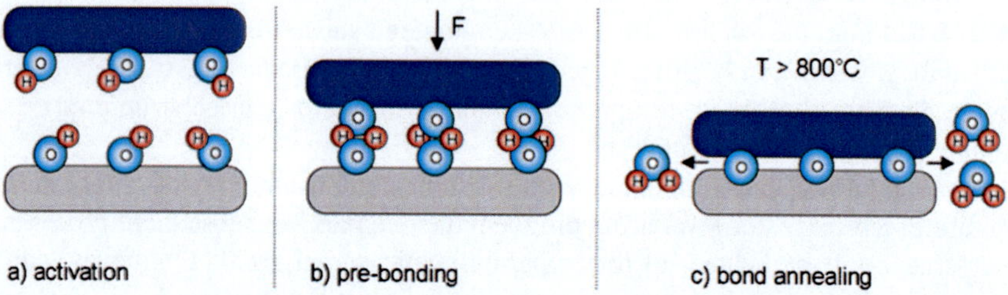

Fig. 3. Chemical effects during high temperature direct bonding

be performed immediately prior to bonding, the wafer processing must be optimized for low roughness. From yield considerations, the whole surface, including the surface area of the wafer edge must be bondable. Therefore, the edge treatment (edge resist removal, edge grinding) and wafer handling (scratches) should also not damage the surface to bond. An edge exclusion zone of about 5–7 mm on 8-in. wafers, which is now state of the art for SOI wafers, is too large for expensive applications or it effects the wafer handling, resulting in process problems. Therefore, special measures of edge engineering are importantly needed and possible to reduce the not bonded wafer edge zone [6]. During the high temperature annealing of bonded wafers with oxide or other layers at the bonding interface, the introduction of stress into the stack by these layers has to be taken into account. In order to minimize stress and stress relaxation effects during wafer processing, the annealing temperature has to be adapted to comply with subsequent high temperature processes such as oxidations and diffusions. Another important feature is that the bonded oxide interface is not as dense as thermally grown oxide, so for example, it has an increased etch rate in HF. This is important to know in order to prevent under cutting of SOI structures or unwanted release (Fig. 4). The density of this bond oxide can be increased by O2-plasma treatment during the activation of the wafers.

3 Wafer Level Capping and Packaging

For wafer-level packaging of typical MEMS and micro- fluidic structures, low temperature and CMOS compatible bond technologies are required. The nearly-completed processed wafers having structures to be capped also contain metal features which must not be degraded by the bonding and activation process. Furthermore, often, e.g. at surface micro machined structures, wet activation is prohibited in order to prevent sticking. The cap wafers are normally specially structured (cavities, through-holes at bond pad areas,

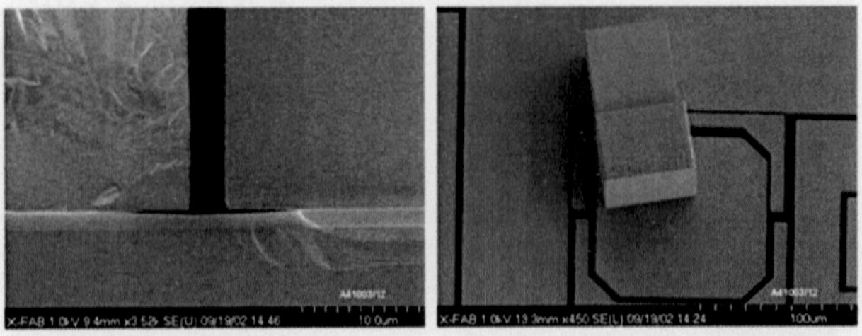

a) under etching at bond oxide   b) released silicon due to under etching

Fig. 4. Under cutting effects, related to fast etching at the bond interface

metal electrodes). In order to contact the electrical structures under the cap, lead-through ducts (at best, metal lines) are required in the bond interface. From the wafer processing, the bond areas are very often rough and not direct-bondable. Additional small bond areas and selective bonding (prevention of bonding movable parts) are required. From the application side, a mechanically strong hermetically tight, reliable bond is usually demanded. From all these aspects, several cap wafer bonding technologies, which will be described below, were developed during the past years.

3.1 Anodic Wafer Bonding

Anodic bonding of glass to silicon wafers is a widely used technique in microsystems technology. This is a relatively simple and safe process, in which a high bond strength and hermetic sealing can be achieved. Typical applications of anodic bonding are sensor encapsulation and sealing of microfluidic systems. For anodic bonding, one of the two wafers to bond must be made of a sodium-containing glass which has its thermal expansion coefficient adjusted to match that of silicon, for example SCHOTT Borofloat33, Corning Pyrex #7740, or Hoya SD2. A bonding stack of one glass and one silicon wafer is heated up to a temperature of 300–500 °C. At these temperatures the sodium ions of the glass become mobile. By applying a negative voltage of some hundred volts to the glass wafers, the sodium ions drift away from the bonding interface, so that a depletion zone is created in the glass close to the silicon, in which most of the bonding voltage is dropped. As a result of the strong electrostatic field, the wafers are pressed together at the atomic level, and, by field-assisted oxygen diffusion and anodic oxidation, oxygen bonds are formed between the glass and silicon. Figure 5 illustrates this process.

The anodic bonding process can be monitored easily from the bonding current. After applying the bonding voltage, the current increases due to the movement of the sodium ions. Then, during the depletion of the glass near the bonding interface, the current decreases again. End point limits for the finished bonding process can be defined as a percentage value of the maximum current, a time, or an amount of charge.

Due to its physics, the anodic bonding is a very safe and reproducible process, producing strong, hermetic bonding. The following layers deposited on silicon wafers can be bonded using this technique: single-crystal and polycrystalline silicon, thermal oxides, CVD oxides, metals and others. Related to the physics of anodic bonding, particularly the high voltage and the movement of mobile sodium ions, the following special protection methods for electrical structures (e.g. piezo-resistors) are required: Well-grounded

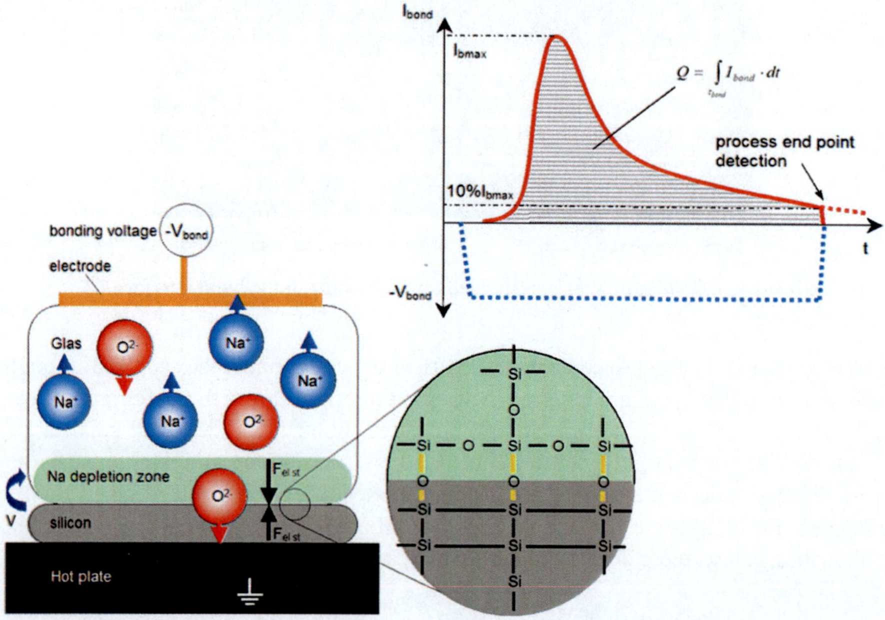

Fig. 5. Principle of anodic bonding and anodic oxidation at the bond interface, including voltage and electrical current over time diagram of a typical anodic bonding process

bonding areas to minimize the field over the electrical structures. The metallization layer can be used for this. To protect the electrical structures against sodium ions, special passivation techniques are required [7].

In order to distribute the electrical bonding field over the wafer area, electrodes of comparable size to the wafer are required. For this, either electrode plates can be installed fixed in the bonder, or electrode wafers can be added to the bonding stack. The use of metallized glass electrodes provides the following advantages:

1. The glass electrode can take up sodium which diffuses out of the glass wafer during bonding. This prevents sodium residuals which cannot normally be removed by cleaning (Fig. 6).
2. If the glass wafer to be bonded has through- wafer holes or slits (perforation), electrical discharges between the silicon wafers and conductive electrodes are likely to occur through the perforations, but can be prevented by using glass electrodes.

The glass used for the anodic bonding has a thermal expansion coefficient (TEC) close to silicon, but nearly equal only at around room temperature. Deviation between glass and silicon can occur as a result of the significant non-linearity of the silicon TEC. At about 300 °C, the expansion of glass and silicon is still similar [8], so that bonding at this temperature gives a stress-free bond after cooling to room temperature. In other cases, the relationship of the TECs of silicon and glass can be used for stress compensation and management in multi-stacked wafers. Anodic glass wafer bonding is

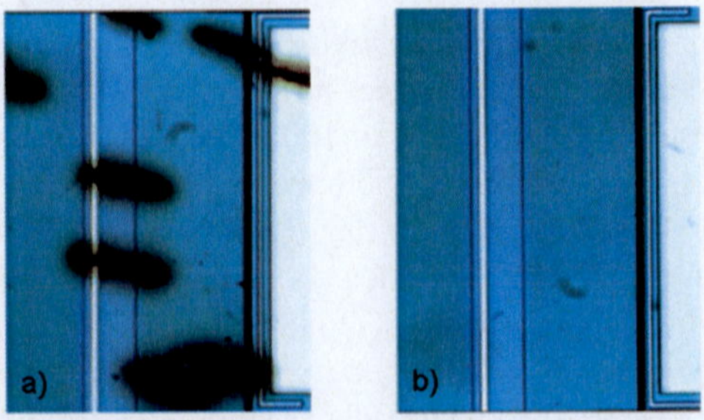

Fig. 6. Prevention of sodium residuals a) at anodic bonded glass wafers by using glass electrodes b)

a well understood and established process, and can by that easily to be adapted to new applications in new technologies.

3.2 Glass Frit Bonding

For glass frit bonding a low melting point glass is used to join two or more wafers. This type of glass is commercially available (e.g. from Ferro Corporation) as a paste consisting of glass powder, organic binders and solvents. In the case of the widely used glass Paste FX 11-036 [9], the active melting glass compound is a lead-silicate glass with a wetting temperature of 425–450 °C, which allows the bonding of wafers with aluminium structures. This glass is non-crystallising and contains zinc oxide as a wetting agent – both these properties are advantageous for good bonding. By adding high-melting barium-silicate filler particles, the thermal expansion behaviour of the glass frit is adjusted to match that of silicon to reduce thermo-mechanical stress at the bonding interface. Meanwhile lead-free glass frit materials with processing temperatures of about 420 °C are in development and evaluation [10]

The glass frit bonding process consists of three main steps: screen printing, thermal conditioning of the glass paste, and the thermo-compressive bonding itself. These steps are shown principally in Fig. 7 and described in the following.

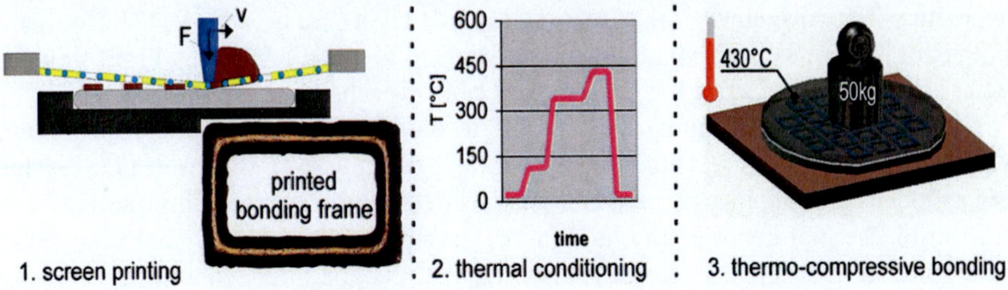

Fig. 7. Principle process flow of glass frit bonding

Using screen printing, the paste can be deposited and structured in one step. The mesh openings are defined by the maximum paste particle size of 15 μm. The mesh thickness relates the printed structure maximum; 30 μm is recommended by the material manufacturer. The resulting bond frames have a minimum width of 190 μm, and a minimum distance of separated structures of 100 μm. To ensure a uniform glass thickness, all structures should have the same width, because the thickness increases with the print structure width. Because screen printing does not give a particularly high accuracy of structural alignment, an area significantly larger than the required bond frame is necessary, and in addition, concave corners require more space. On the other hand, very precise screen printing is possible if the bond frame is limited by a cavity on both sides. For this, the screen opening has to be broader than the bond frame area defined in the silicon by the limiting cavities prior to bonding, a multistep thermal conditioning (Fig. 8) is necessary in order to transform the paste into a real glass. During this pre-conditioning the solvent is burned out and the glass is pre-melted without internal voids.

The bonding is a thermal process supported by slight pressure. By heating up the glass to the wetting temperature, its viscosity decreases to the point where it wets the bond surface. Material from the surface layer fuses into the glass at the atomic level, so that a strong bond is formed during cooling. The bonding temperature is the critical bonding parameter because it must be high enough to guarantee good wetting of the bond surface with the glass as the initial process for bond formation. The bond pressure gives only some support for the wetting, and equalises wafer geometry inadmissibility, such as bow and warp. The cooling of the bonded wafer pair is only critical at higher temperatures where the bond is finally formed. However, to prevent thermal cracking of the wafers or the bonded interface by thermal shocks, the wafer stack has to be cooled below 200 °C before removing it from the bonding chamber.

By using a low melting point glass frit, it is possible to bond nearly all the surface materials used in silicon microsystems: Silicon: (single and polycrystalline of any doping level) insulators and passivations (thermal or CVD silicon oxide, LP- and PE-CVD silicon nitride), metals (aluminium, aluminium-silicon, titanium, titanium-nitride, titanium-tungsten) special layers (polyimide, indium-tin oxide). Bonding wafers of different materials, such as silicon and Pyrex-type glass, is also possible if the thermal expansion coefficients are in the same range.

At sufficient screen printing quality, sealing of vacuum in the sensor cavities is possible, but the resulting pressure is in the range of several mbar (measured by resonant structures). This is caused by out-gassing of organic residuals from the glass into the sensor cavities. To realise lower cavity pressures, getter layers are necessary [11]. Hermetic bonding of surface steps of up to few micrometers (e.g. metallic electrical lead-through ducts) is also possible, since the melted glass can cover these steps (Fig. 8c).

Glass frit bonding is a universally usable technique for wafer level encapsulation, even for high performance applications such as gyroscopes, as shown in Fig. 9. This very safe and reproducible process consists of three main steps: screen printing of the glass paste, thermal conditioning, and thermo-compressive bonding. Characterisation of the bonded wafers and glass frit has shown that this bonding technique gives many advantages in practical use.

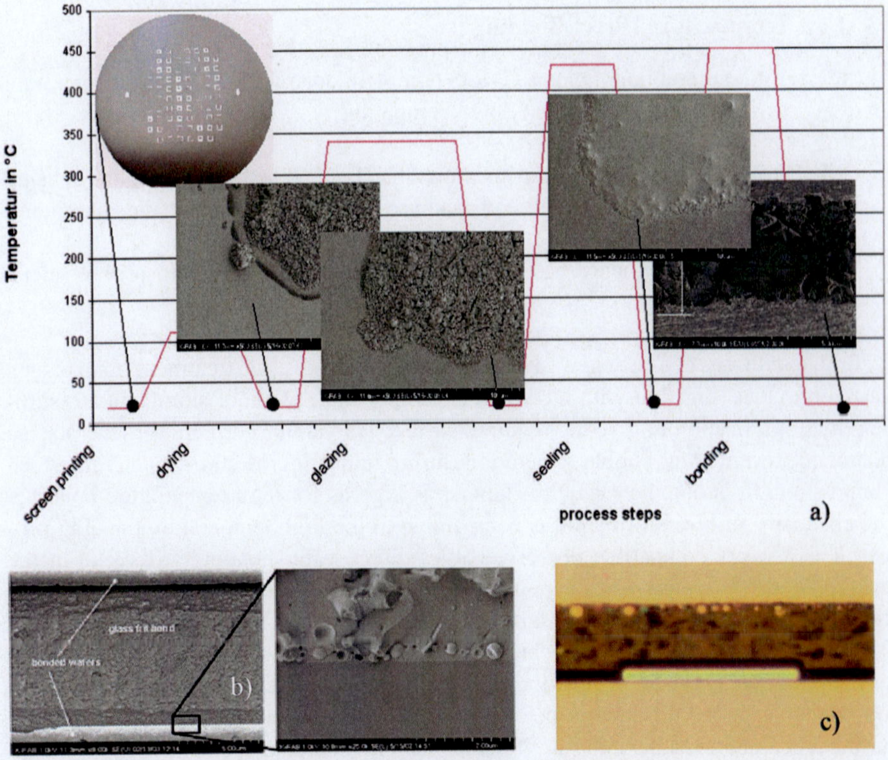

Fig. 8. a) thermal conditioning glass frit paste and b) glass frit bond interface with c) sealed metal line as electrical signal lead through

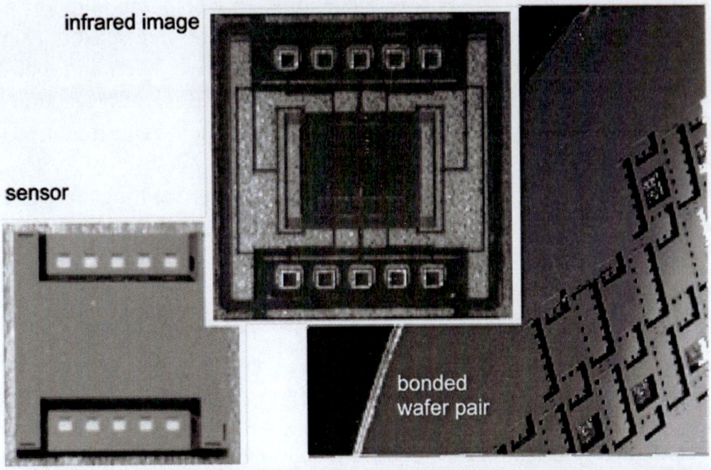

Fig. 9. Glass frit bonded gyroscopes as example application of glass frit bonding

3.3 Low Temperature Direct Bonding

Low temperature direct bonding is a CMOS compatible variant of the high temperature bonding described above. To enable CMOS compatibility, two conditions are necessary:

1. CMOS compatible activation and cleaning, the classical activation techniques using RCA or other acids are prohibited because these would attack the aluminium metallisation.
2. The annealing temperature has to be lowered to about 400 °C, to prevent metal alloying.

Both conditions can be achieved by oxygen plasma activation treatment. The oxygen plasma provides, together with DI-Water rinsing or multiple-hour storage in air, sufficient hydrophilisation to give the required surface activation. Furthermore, the plasma treatment prevents any bubble generation during annealing at 200–400 °C, in which a bond strength is obtained which is almost as high as for high temperature bonding. The chemistry of bond formation is comparable to the high temperature bonding process. This CMOS compatible process enables direct wafer bonding at the end of the line (back-end) for wafer level capping applications. However, with this process there are high demands on the bonding surface quality. For safe bonding, an R_a value below 1–2 nm is required, which can be achieved only with polished surfaces. Any plasma processing, such as CVD layer deposition and etching, will roughen the surface and render it non-bondable. To use this low temperature bonding in industrial wafer processing, where plasma processes are state of the art, it is necessary to protect the bond interface from the beginning. At the end of the process, prior to bonding, these layers must be removed by wet etching to obtain bondable areas. During this removal the active structures have to be protected. Low temperature bonding process has been used for support-wafer bonding of bulk micro-machined pressure sensors. Depending on whether the support wafer was perforated or not, the process gave relative or absolute pressure sensors (Fig. 11). In this case, a back-side etching mask was used as a protective layer, and then removed before bonding using a single-side spin etch processor. Sufficient protection was also found possible using a thermal oxide on support wafers which were structured (hole) by powder blasting.

Because low temperature direct bonding is so sensitive to the roughness of the bonding surface, it is rarely used as a back-end capping process. A safe, high-yielding process can only be achieved if low roughness is obtained after the main wafer processing (protective layer or CMP). To achieve a sufficiently low roughness at the end of a wafer process always requires extra effort and a special process flow (Fig. 10).

3.4 Other Wafer Bonding Technologies for MEMS Wafer Level Packaging

In recent years several more wafer bonding technologies found their way from basic process development into the MEMS Applications. Adhesive wafer bonding using thin span on layers are used for examples for highly sophisticated MEMS CMOS integration like bolometers and switches [12] while thick dry film layer laminated and structured on a single wafer can form microfluidic channels and can also act as a bonding layer to

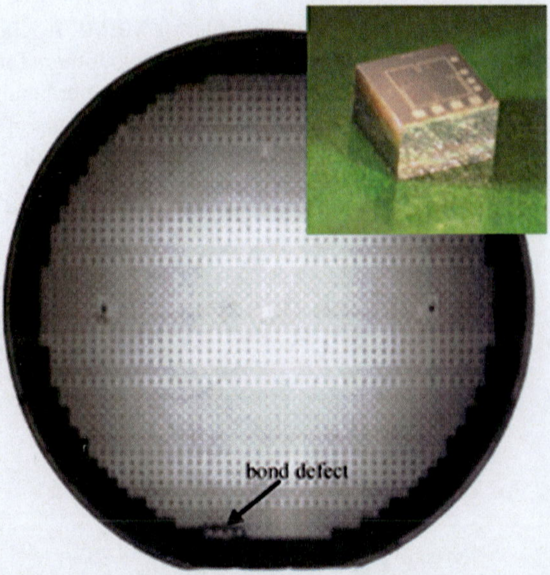

Fig. 10. Example of low temperature direct bonding absolute pressure sensor fabricated by backside sealing of a relative pressure sensor wafer

bond a lid wafer as cap to the microfluidics [13]. For hermetic sealing of MEMS sensor structures metal-based bonding technologies are emerging since they are providing compared to glass frit bonding lower vacuum levels at smaller bond frame widths. Eutectic bonding especially as Al-Ge Bonding in a CMOS compatible version [14] solder boding (the metal solder can be electroplated or screen printed) [15] and Metal-Metal-thermo compression bonding [16] should be named here. Regarding process integration and application related aspects they have specific advantages and drawbacks. Very interesting is the surface activated bonding [17] which is based on an ion bombardment of the bond surface, cleaning them from oxides and contaminations, leaving a very activated surface allowing. Strong and reliable wafer bond at room temperature without a need of later annealing.

4 Summary

For wafer bonding, which is a very important MEMS manufacturing step, several well-known, technologies are available. The most frequently used of these, high and low temperature direct bonding, and glass frit and anodic bonding, have been described in this paper. It has been shown that all of these techniques give many possibilities for process integration and application. However, there are also different challenges in the utilization of each technology for safe, reproducible MEMS fabrication. Figure 11 classifies the technologies introduced here, as well as other important wafer bonding technologies, such as metal interlayer or adhesive bonding, in terms of bonding temperature, bond strength, hermeticity and CMOS compatibility.

All the technologies discussed here are suitable for MEMS Applications and their production, they each have particular advantages which favour them for certain applications. Table 1 gives some general suggestions of how to use wafer bonding technologies in different technological fields. Importantly is beside the suitability of a wafer bonding process is the process integration in to the complete MEMS technology, bonding temperatures, processing of the bonding materials and surface qualities are must considered very carefully.

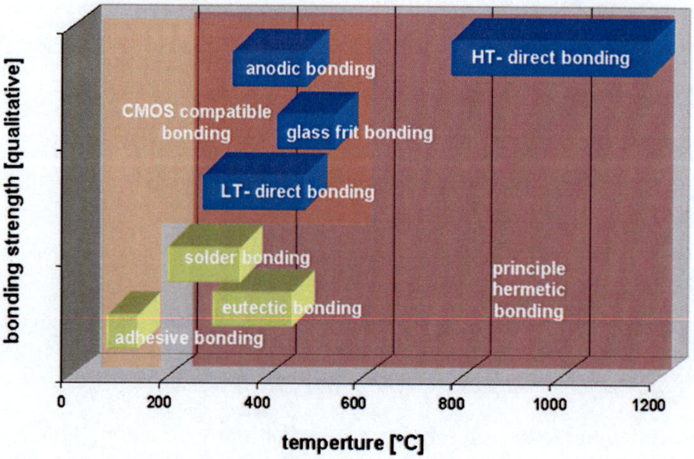

Fig. 11. Classification of different bonding technologies in terms of bonding temperature, bond strength, hermeticity and CMOS compatibility.

Wafer bonding technologies are still in a phase of intensive research and development. The technologies described in this paper are continuously improved regarding both their bond interface properties and their process integration aspects to use them for even more and advanced applications. New technologies are emerging in research like reactive bonding where the heat to form the bond is generated by chemical reaction of metal layers directly on the wafers [18]. Since deposition and structuring of the bonding frame materials as described for the glass frit bonding is difficult and high in effort, additive printing process are becoming interesting for wafer bonding applications [19]. In industry the so-called hybrid bonding is a very important development topic. This kind of bonding is allowing by utilizing modern chemical mechanical polishing methods the simultaneous bonding of oxide and metal surface to form both a strong mechanical bond and electrical contacts in one and the same process [20]. So finally, it can be concluded wafers boding is and stays and a very important method for micro system technologies.

Table 1. Suitability of wafer bonding processes for applications ++ well suitable + suitable (+) conditional sutable - not recommended – not suitable

Wafer Bonding Process	Engineered Substrates	MEMS Capping	MEMS CMOS Integration	Micro Fluidics
High temperature direct bonding	++	–	–	+
Low temperature direct bonding	+	+	+	+
Glass frit wafer bonding	–	++	++	–
Eutectic bonding	–	++	++	–
Solder bonding	–	++	++	–
Metal direct bonding	–	++	++	–
Adhesive bonding	–	(+)	+	++
Surface activated bonding	++	+	+	++

References

1. Wiemer, M.: Technologieentwicklung für Beschleunigungssensoren und Drehratensensoren unter Nutzung von Waferbondverfahren, Dissertation April 1999 Chemnitz
2. Knechtel, R.: Single Crystalline Silicon Based Surface Micromachining for High Precision Inertial Sensors - Technology and Design for Reliability, Springer Microsystem Technologies, vol. 16, https://doi.org/10.1007/s00542-010
3. Smith, A., Moore, J., Hosse, B.: A chemical and thermal resistant wafer bonding adhesive simplifying wafer backside processing. In: CS MANTECH Conference, 24–27 April 2006, Vancouver, British Columbia, Canada (2006)
4. Info Sheet: 1…3-Axis Surface Micromachining Process for Inertial Sensors Process Family XMB10, X-FAB MEMS Foundry GmbH
5. Wiemer, M., Zimmermann, S., Zhao, Q.T., Trui, B., Kaufmann, C., Mantl, S., Dudek, V., Gessner, T.: Fabrication Of Soi Substrates With Buried Silicide Layers For Bicmos Applications, Semiconductor Wafer bonding VIII, Science, Technology, and Applications (2005)
6. Knechtel, R., Schwarz, U., Dempwolf, S., Nevin, A., Klingner, H., Lindemann, G., Schikowski, M.: The Role of Wafer Edge in Wafer Bonding Technologies, Semiconductor Wafer Bonding 16; PRiME 2020, Honolulu, HI, 4–9 October 2020 (2020)
7. Freywald, K., Knechtel, R.: EP000001270507B1 Passivation of anodic bonding regions in microelectromechanical systems
8. Harz, M.: Untersuchung thermo-mechanischer Spannungen beim anodischen Bonden von TEMPAX-Glas und Silizium, Dissertation 1996, TU-Dresden
9. MEMS & Sensor Materials 11-036 Sealing glass, Verarbeitungspezifikation FERRO Electronic Materials Santa Babara, Ca, Rev. 1200 (2002)
10. Tilli, M., Paulasto-Krockel, M., Petzold, M., Theuss, H., Motooka, T., Lindroo, V.: Handbook of Silicon Based MEMS Materials and Technologies, 3rd edn., p. 605f (2020)

11. Moraja, M., Amiotti, M., Longoni, G.: Patterned Getter Film Wafers for Wafer Level Packaging of MEMS. Micro Syst. Technol., München (2003)
12. Niklaus, F., et al.: Low-temperature full wafer adhesive bonding. J. Micromech. Microeng. **11**(2) (2001)
13. Karl, W., Schikowski, M., Thon, J.-E., Knechtel, R.: Adhesive wafer bonding for CMOS based lab-on-a-chip devices Published 3 Jan. 2020 The Japan Society of Applied Physics (2020)
14. Tavassolizadeh, A., Kulkarni, A., Zellmer, M., Thon, J.-E., Maes, B., Buggenhout, C.V.: Al/ge Eutectic Wafer Bonding for MEMS/CMOS Vacuum Packaging, Proceeding Wafer Bond 2019 (2019)
15. Sparks, D., et al.: Wafer-to-wafer bonding of nonplanarized MEMS surfaces using solder. J. Micromech. Microeng. **11**(6) (2001)
16. Dempwolf, S., Knechtel, R.: Standard bond test wafers for evaluating wafer bonding technologies. In: Proceeding Wafer Bond 2009 (2009)
17. Suga, T., Mu, F.: Surface activated bonding method for low temperature bonding. In: 2018 7th Electronic System-Integration Technology Conference (ESTC), Dresden, pp. 1–4 (2018). https://doi.org/10.1109/estc.2018.8546367
18. Bräuner, J., Besser, J., Wiemer, M., Gessner, T.: Room-temperature reactive bonding by using nano scale multilayer systems. In: Solid-State Sensors, Actuators and Microsystems Conference (TRANSDUCERS) (2011). https://doi.org/10.1109/transducers.2011.5969498
19. Wiemer, M., Roscher, F., Seifert; T., Vogel, K., Ogashiwa, T., Gessner, T.: Low temperature thermo compression bonding with printed intermediate bonding layers. In: Semiconductor Wafer Bonding 14 PRiME 2016 (2016). https://doi.org/10.1149/07509.0299ecst
20. Di Cioccio, L., et al.: An overview of patterned metal/dielectric surface bonding: mechanism, alignment and characterization. J. Electrochem. Soci. **158**(6), P81. https://doi.org/10.1149/1.3577596

Conceptual Design of a Microscale Balance Based on Force Compensation

Karin Wedrich[1](✉), Maximilian Darnieder[2], Eric Vierzigmann[1], Alexander Barth[3], Rene Theska[2], and Steffen Strehle[1]

[1] Department of Mechanical Engineering, Microsystems Technology Group, IMN MacroNano®, Technische Universität Ilmenau, Ilmenau, Germany
karin.wedrich@tuilmenau.de

[2] Department of Mechanical Engineering, Precision Engineering Group, Institute for Design and Precision Engineering, Technische Universität Ilmenau, Ilmenau, Germany

[3] Department of Mechanical Engineering, Institute of Process Measurement and Sensor Technology, Technische Universität Ilmenau, Ilmenau, Germany

Abstract. Macroscopic electromagnetic force compensation (EMFC) balances are well established but were not yet demonstrated within microsystems. Hence, in this paper, the concept and the design of a micro fabricated force compensation balance is presented. The implemented concentrated compliance mechanism in form of flexure hinges enables motion with high precision, which is combined with a force compensation mechanism. The concept of force compensation promises a high measurement range, which is expected to be up to 0.5 mN, while still enabling a high resolution of less than 8 nN. The developed dynamic model of the miniaturized balance is used for the design of a PID-controller strategy. Here, continuous and time-discrete controller approaches are compared. The time-discrete controller with realistic delay times, leads to an accuracy of the controller, which is better than the expected accuracy of the integrated capacitive position sensor.

Keywords: Force compensation balance · MEMS · Force sensor · PID controller

1 Introduction

Miniaturized sensors for ultra-small mass and force sensing are required for various applications ranging from nanoscale metrology to microbiology and micro biophysics [1–3]. To measure small forces in the range of Nanonewton to Millinewton with the same microsystem, mostly microcantilevers are currently the preferred transducer [3–5].

A significant advantage that fostered the use of microcantilevers is their straightforward operation principle accompanied by the ease of microfabrication at relatively low cost. Microcantilevers are frequently employed in atomic force microscopy (AFM), where the mechanical cantilever deflection is recorded via an optical or a piezoresistive readout strategy. Nevertheless, there are several aspects of microcantilevers that must be critically reviewed. The optical readout approach for instance requires accurate

L. Zentner and S. Strehle (Eds.): MAMM 2020, MMS 96, pp. 103–114, 2021.
https://doi.org/10.1007/978-3-030-61652-6_9

adjustment of the optical parts to avoid impacts on the measurement accuracy. The surface area of the microcantilever that is designated for the optical readout must be free of particles and other surface features that impair the optical signal. Additionally, the measurement range of microcantilevers is intrinsically restricted [6, 7] and the relation between the acting input force and microcantilever deflection is non-linear, which gains in particular importance with progressing beam deflection. Hence, microcantilevers are ideal tools for force sensing if the expected force range is relatively small. However, if an extended force measurement range is required that simultaneously offers a constant accuracy or force sensing resolution within the entire operation, micro sensor concepts are required that are based on principles that do not depend on mechanical deflection. The so-called compensation principle, which is realized in electromagnetic force compensated (EMFC) balances, provides this possibility [8–10]. In brief, an external force produces a deflection that is measured and compensated by a closed loop control system and for small forces, an electrostatic actuator is used also in the macroscopic [11, 12]. The principle deflection compensation by feedback control is also common in microsystems mostly in accelerometers [13–15] but is not yet established for high precision force or mass micro sensors. With emphasis to AFMs, microcantilevers compensated by closed loop control with sensing and actuating piezoelectric layer were demonstrated in [16] and another compensation system with a lever with two torsion bar springs are presented in [17], which reached a measurement range of $\pm 32\ \mu N$ with a resolution of 1.96 nN. A force sensor presented in [18] offered a measurement range of 1.5 mN and a resolution of 7.8 nN by utilizing a microscale comb drive system and electro-thermal sensing elements. Hence, the overall approach of this paper is supported.

In this paper, a concept for a first micro fabricated force compensation balance is presented. Its micromechanical system is based on the well-established mechanism concept of macroscopic electromagnetic force compensation balances but uses electrostatic principles, which are pronounced due to scaling effects. It promises the highest measuring range of micro fabricated force sensors with a constant resolution of just a few nanonewtons. The paper comprises the system design accompanied by the calculation of the measurement range and the expected resolution of the miniaturized compensation scale as well as the dynamic model to design a PID controller to control the balance.

2 System Design

Inspired by macroscopic EMFC balances, the micromechanical system was designed based on concentrated compliance in form of flexure hinges to achieve a precise motion. The mechanical system can be divided into the functional elements *linear guiding system* and *transmission lever*. The linear guiding system is a parallelogram linkage including upper lever (1), lower lever (2) and shuttle (3) and four flexure hinges (A–D), as shown in Fig. 1. The external force F_{load} or mass acts on the shuttle (3). The transmission lever (5) is a beam suspended by flexure hinge (G). This functional element is coupled over two flexure hinges (E, F) to the coupling element, which is again coupled to the shuttle.

If an external force is applied to the shuttle, the shuttle is guided vertically with the (quasi-)linear guiding system. Forces in x-direction are suppressed but out-of-plane forces can have an influence. However, the stiffness in this direction is much larger than

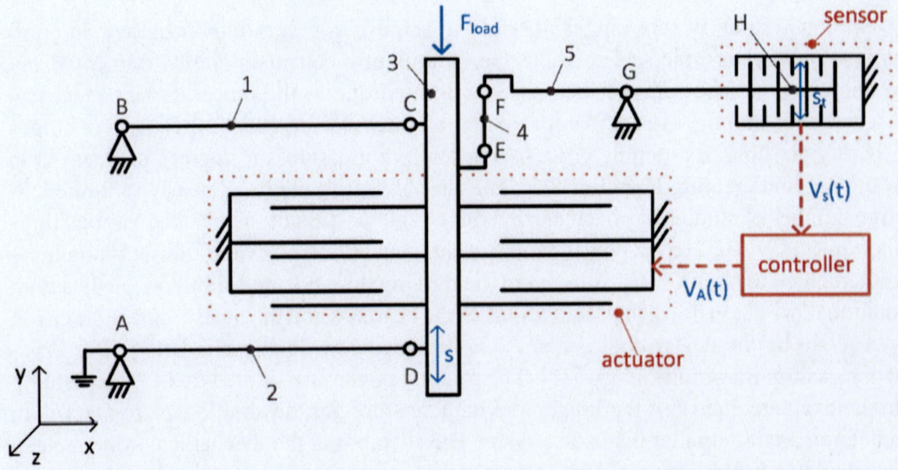

Fig. 1. Illustration of the working principle of the designed microscale compensation balance: 1- upper lever, 2- lower lever, 3- shuttle, 4- coupling element, 5- transmission lever

in the *y*-direction. The deflection *s* under the loading force F_{load} depends on the stiffness of the balance *c* according to

$$F_{load} = cs \tag{1}$$

This deflection is passed to the transmission lever and is increased by the transmission ratio *tr,* which depends on the position of the flexure hinge G, while l_{GH} is the length of the lever from the hinge G to position H and l_{FG} is the length of the lever from the hinge F to hinge G.

$$tr = \frac{l_{GH}}{l_{FG}} \tag{2}$$

The position sensor detects the deflection of the transmission lever s_t. Here, an integrated comb drive capacitive sensor is used to avoid influences on the sensor signal e.g. due to temperature changes, which is for instance the case if a thermoelectric sensing strategy is used. An additional circuit converts the capacitance to an electrical signal (voltage). A controller uses the voltage signal of the transducer to determine the voltage for the electrostatic actuator, which includes the information of the loaded mass. The actuator is realized by a comb drive with asymmetric structure to move in transverse direction with *n* fingers. By neglecting leakage fields, the force can be approximated with:

$$F_{load} = \frac{1}{2}V^2\frac{dC(x)}{dx} = \frac{1}{2}\varepsilon_0\varepsilon_r AV^2\left(\frac{n}{(d_{10}-s)^2} - \frac{n}{(d_{20}+s)^2}\right) \tag{3}$$

While *V* is the potential difference between the fingers, *A* is the area of the actuator fingers on the sidewalls, d_{10} is the smaller gap between actuator fingers and d_{20} is the larger gap in non-deflected position. This kind of actuator leads to large forces, but the forces are dependent on the changing gap between the actuator fingers.

In this system, flexure hinges are used to achieve a high motion accuracy. In comparison to macroscopic scales where the compliant mechanism is also fabricated out of one element. The form of the hinges is limited due to the processes like electrical discharge machining, even at macroscopic scales a perfect shape of notch flexure hinges is hardly possible. By etching processes in microfabrication, the degrees of freedom in designing and creating lateral or 2.5 dimensional morphologies is nearly unlimited. A large number of studies has been carried out to analyze flexure hinges and various morphologies [19, 20]. Here a power-function shape based on a seven-order polynomial is implemented to obtain a low stiffness in the moving direction and simultaneously a high rotational precision, too. The calculations of the stiffness and the measurement precision were done by the design tool *detasFLEX*, developed by Technische Universität Ilmenau, based on Bernoulli-theory [21]. The rotation angles are limited to 3–7°, depending on the exact geometry of the hinge. Mechanical stops for the shuttle ensure maximum deflection angles smaller than one degree. The stiffness of the mechanical system c can be calculated [22] by:

$$c = \frac{1}{l_{FG}^2}\left(4c_{hinge}\frac{l_{FG}}{l_{AD}}^2 + 2c_{hinge}\right) \tag{4}$$

For the force resolution of the measurement system, the spatial resolution of the sensor is crucial. The theoretical force resolution can be calculated by:

$$\Delta F = c\Delta s_t \frac{1}{t_r} \tag{5}$$

To define the optimal parameters for the force compensation balance, Eq. (5) is analyzed with the stiffness equation in (4). The stiffness of the single hinges has a linear and therefore the most pronounced influence on the resolution. By increasing the length l_{AD} or l_{GH}, the resolution reaches a boundary value. By fixing the length $l_{GH,}$ for the length l_{FG} there is an optimum.

Table 1. Parameters of the force compensation balance.

Parameter	Value	Unit	Parameter	Value	Unit
l_{AD}	1700	µm	$F_{\max}$	0,5	mN
l_{FG}	1000	µm	ΔF	11	nN
l_{GH}	4000	µm	size	1,05 × 0,5	cm^2
c_{hinge}	13.2	Nµm	k_{SF}	0.27	mNs/m
c	44.6	N/m	k_{Sq}	1.7	µNs/m

To account for the maximum force capacity of 0.5 mN, 108 actuator fingers are here required. Notably, the maximum force can be readily increased by increasing the numbers of actuator fingers in the comb system. The comb-drive sensor uses in total 125 fingers arranged in a differential configuration. For a deflection s_t of 1 nm a change in capacitance of 34 aF must be resolved. Therefore, one line of sensor fingers is used for positive capacitive changes and one line of sensor fingers is used for negative capacitive changes to avoid additional parasitic capacitance.

Dynamic Model of the System

For the description of the dynamic behavior of the system, a model is developed, which is based on the torque around the frame fixed hinge G. In the simplified illustration (Fig. 2), the effecting forces are shown. In the illustration, the balance is oriented in a way, such that the gravitational acceleration is acting in the negative *y*-direction.

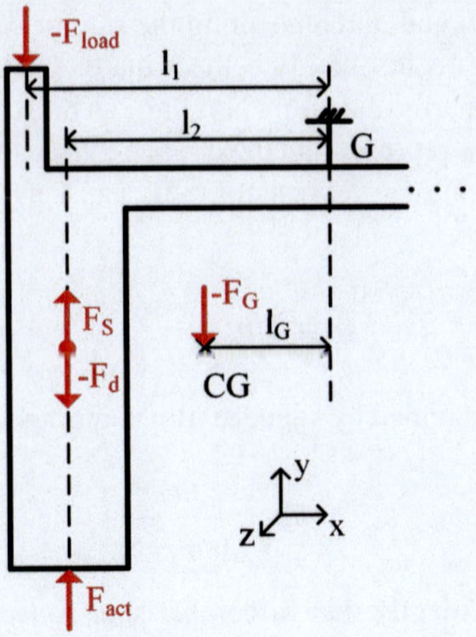

Fig. 2. Simplified illustration of the mechanical part of the force compensation balance with all considered forces required for the dynamic model.

For this model, just the rotation around the *z*-axis is considered. Due to the law of conservations of angular momentum, the intrinsic angular momentum D_G around the frame fixed hinge G is given by:

$$\dot{\overrightarrow{D_G}} = J\ddot{\varphi} = \overrightarrow{M_G} \tag{6}$$

With the moment of inertia J, here J_{ZZ} around G, the angular acceleration $\ddot{\varphi}$, here $\ddot{\varphi}_z$ and the torque M_G. For the deflection s of the shuttle in *y*-direction, the small angle approximation can be used leading to the following equation:

$$s = l_2 \sin(\varphi) \approx l_2\,\varphi \tag{7}$$

This means to the torque M_G following:

$$J\frac{\ddot{s}}{l_2} = \overrightarrow{M_G} \tag{8}$$

By considering the actuation force, the elastic force and the damping force, the weight force and the loading force, the torque M_G can be also described by:

$$\overrightarrow{M_G} = \overrightarrow{F_{act}} l_2 + \overrightarrow{F_S} l_2 + \overrightarrow{F_d} l_2 + \overrightarrow{F_G} l_G + \overrightarrow{F_{load}} l_1 \tag{9}$$

Hence, the following equation describes the dynamic behavior of the system

$$\ddot{s} = \frac{l_2}{J}\left(\frac{1}{2} * \varepsilon_0 \varepsilon_r A V^2 \left(\frac{n_1}{(d_{10} - s)^2} - \frac{n_2}{(d_{20} + s)^2}\right) l_2 - c s l_2 - k \dot{s} l_2 - g(l_G M + l_1 m)\right) \tag{10}$$

with M representing the mass of the moving parts, transmission level, shuttle and half of the upper and lower level and m representing the sample mass.

In this system, damping occurs mainly in the actuator and the transducer element. On the transducer element, primarily slide film damping can be expected, which is calculated here as per ref. [23] with η_{air} representing the dynamic viscosity of air, A_{SW} representing the area of the sidewalls and d_{SW} representing the gap between the sidewalls and the actuation fingers:

$$k_{SF} = 2n\mu_{air} \frac{A_{SW}}{d_{SW}} \tag{11}$$

The actuator is mainly damped by squeeze film damping, calculated as per ref. [24]:

$$k_{Sq} = n\mu_{air} \frac{h^3 b}{(d_0 - z)^3} \tag{12}$$

The equations show evidently that in comparison, squeeze film damping is three orders of magnitude bigger than slide film damping by the used parameters see Table 1. Hence, only squeeze film damping will be considered further in this model.

3 Fabrication

To facilitate a future microtechnological system fabrication, the overall workflow and the material choices as shown in Fig. 3 were selected in alignment with established thin film processes based on silicon-on-insulator (SOI) substrates (1). First, the contact pads on top of the device layer will be deposited. The contact pads are intended to consist of a metal stack of first 400 nm aluminum, 10 nm chromium followed by a 200 nm gold film. The entire metal stack is patterned via a lift-off process (2). The workflow requires to use a HF vapor at the end of the device fabrication. Therefore, chromium is used as a separation layer between aluminum and precious gold to avoid aluminum corrosion. The metal will be annealed after lift-off patterning to yield an aluminum-silicide formation, which shall lower the energetic contact barrier and thus, yield a low-resistance ohmic contact behavior.

The functional structures on the SOI device layer (100 μm) are subsequently patterned employing a deep reactive ion etch process (3). The SOI handle layer (400 μm) must be partially removed to ensure an unrestricted movement of the moving lever. To

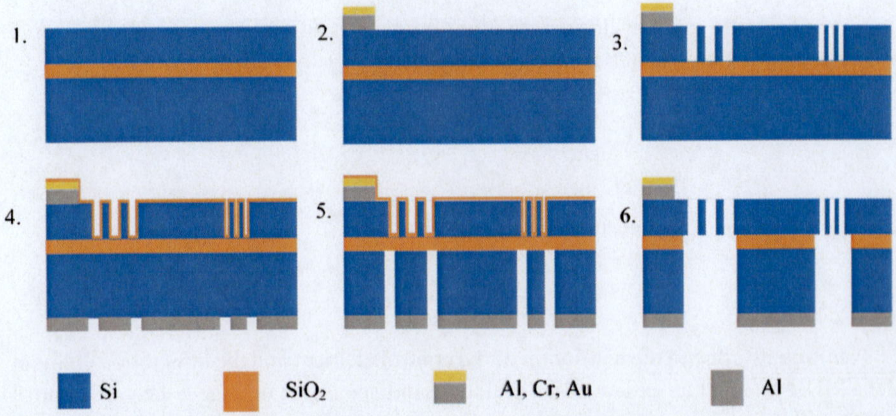

Fig. 3. Fabrication process flow: 1: SOI- Substrate, 2: deposited contact pads, 3 structured device layer, 4: deposited hard mask for backside etching, 5: etched handle layer, 6: etched box layer

avoid any fractures in the SOI box layer (2 μm) due to different etch-ditch-widths, support structures are considered in the overall design. For the patterning of the handle layer, an aluminum hard mask is implemented (4). However, before the handle layer can be patterned by deep reactive ion etching, a protective SiO_2 layer (deposited by ICPCVD) is required and deposited onto the device layer (5). During the backside etching process, the systems will be separated and finally, the protective SiO_2 layer and the box layer are etched by a dry HF vapor etching process to release the movable system parts (6).

4 A First Approach to Design a PID Controller

The basis for the controller is the dynamic model of the system as described before in Eq. (10). Comparable to macroscopic electromagnetic compensation balances [25], a PID controller is considered. By analyzing Eq. (10), it becomes apparent that the controlled system appears to behave unstable and non-linear, which requires an adequate control mechanism.

4.1 Continuous Control

At first, a continuous controller is developed. Due to the non-linearity of the system, a linearization should be done in minimum one operating point. A common option is a Taylor series. An omittance of Taylor polynomial higher-orders, will lead to inaccuracies and will potentially work only far away from the operating point. Here, a common and more precise alternative is consequently considered, which is given by the feedback linearization [26]. Therefore, it is necessary to substitute all non-linear terms of the system. For these terms, a controller is designed and subsequently, the controlled variable will be re-substituted and transferred to the controlled system. The non-substitute terms will be compensated by the controller, so it is expedient to minimize these terms. In this case, in addition to the non-linear actuation force term, all terms except the load force will be substituted to w(t) (13),

$$v(t) = \frac{l_2}{J}\left(\frac{1}{2} * \varepsilon_0\varepsilon_r A v(t)^2\left(\frac{n_1}{(d_{10} - s(t))^2} - \frac{n_2}{(d_{20} + s(t))^2}\right)l_2 - csl_2 - k\dot{s}l_2 - gl_G M\right) \tag{13}$$

whereof a new controlled system is generated (14):

$$\ddot{s}(t) = w(t) - \frac{l_2}{J}mgl_1 \tag{14}$$

Additionally, due to the restriction of the control element and the integration element of the PID-controller an anti-wind-up method is indispensable. In Fig. 4, the used control circuit is illustrated. In this circuit, two limitations of values are included, the deflection s to ± 5 μm due to mechanical stops in the design and the voltage V from 0 V to + 100 V to be safe because of the disruptive strength of the SiO_2 of the SOI box layer. The maximum voltage will be determined in experiments.

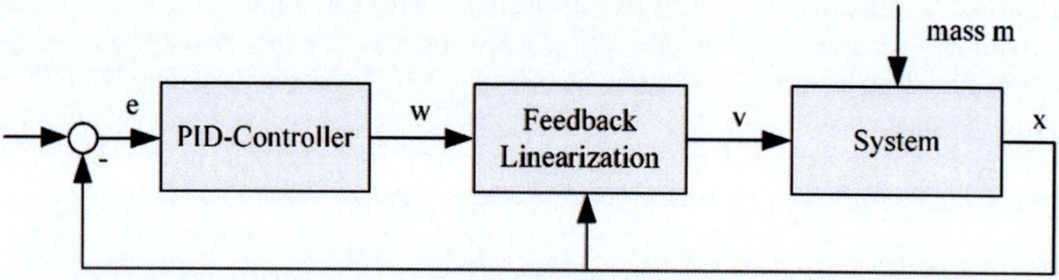

Fig. 4. Illustration of the control circuit of the continuous PID-controller.

4.2 Time-Discrete Control

For a real system with a capacitive-to-digital converter and a voltage source to read-out the sensor and drive the actuator of the force compensation scale, a time discrete controller is necessary. The sampling time of the controller should be slower than the sampling rate of the capacitive-to-digital converter to record the movement of the shuttle and react to it properly. Here, the sampling rate in the discretization block is set to 0.1 ms. To consider delay times due to signal transmission and the dynamic of the voltage amplifier, a time delay element is included into the control circuit with a time constant of 10^{-5} s. Figure 5 shows the control circuit of the time discrete PID-controller (red: discrete signals).

4.3 Comparison of Continuous and Time Discrete PID Controller

The parameter of the PID controller are adapted to each controller and listed in Table 2, while *Kp* represents the proportional gain, *Ki* the integral gain, *Kd* the derivate gain and N the filter constant.

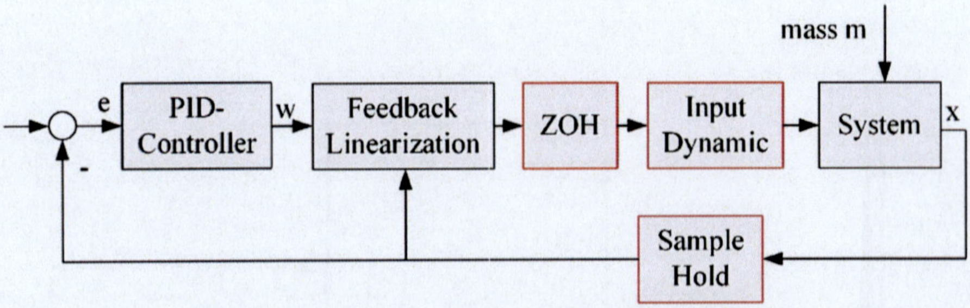

Fig. 5. Control circuit of the time-discrete PID-controller.

Table 2. Parameters of both the controllers.

Parameter	Continuous controller	Time-discrete controller
Kp	20.000.000	30.000.000
Ki	5.000.000.000	16.000.000.000
Kd	7000	9000
N	8.500.000	16.000

In Fig. 6, the simulated deflection of the shuttle by the continuous and discrete controller is shown after applying a maximum deflection to begin the simulation. After 30 ms, a mass of 50 mg is applied and again after 30 ms a mass of 5 mg is applied. For the same scenario in Fig. 7, the simulated voltage for the actuator to balance the lever is shown. In Table 3, the control deviation of both controllers are listed by applying a mass of 5 mg and 50 mg after 20 ms and 50 ms of simulation time.

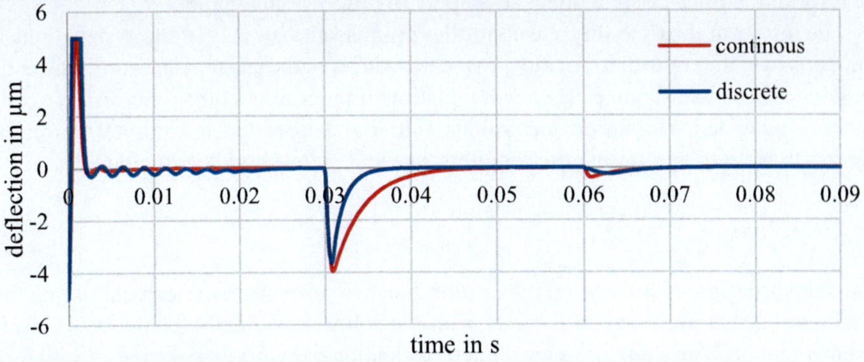

Fig. 6. Simulated reaction of the shuttle by the controller: **initial**: maximum deflection of $-5\ \mu$m; **0.03 s**: mass of 50 mg; **0.06 s**: mass of 5 mg.

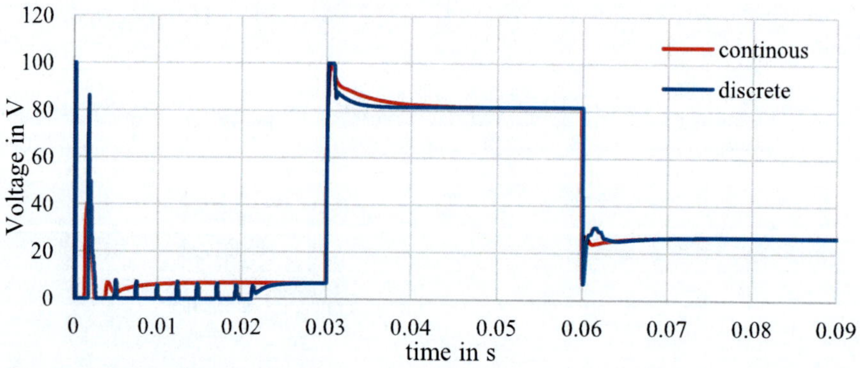

Fig. 7. Simulated voltage for the actuator to balance the lever by controller: **initial**: maximum deflection of −5 μm; **0.03** s: mass of 50 mg; **0.06** s: mass of 5 mg.

Table 3. Control deviation of both the controllers.

Delay Time in s	Continuous		Time-discrete	
	m = 5 mg	m = 50 mg	m = 5 mg	m = 50 mg
0.02	2.02 μg	21.8 μg	0.0004 μg	0.0145 μg
0.05	0.005 μg	0.05 μg	$4.57 \cdot 10^{-13}$ μg	$2.96 \cdot 10^{-11}$ μg

By comparing both controllers, it becomes evident that the same voltage should be applied to the actuator to balance the lever. For the discrete controller, it takes more time to control the shuttle to a stable position by a maximum deflection. It basically swings more. To adjust the mass, which does not lead to a maximum deflection, the discrete controller is more rapid, consequently, the deflection of the shuttle is less than by the continuous controller. In addition, by comparing the control deviation of both controllers the error of the time-discrete one is less up to 10 orders of magnitude.

The results of the time-discrete controller promise an accuracy of the system, which is not linked to the controller, but more to the transducer or the capacitance-to-digital converter and the voltage source. The contemplation of the results of the time-discrete controller gives evidence for the devices voltage source and capacitance-to-digital converter, which can be used by limiting the sampling rate and increasing of the voltage.

5 Conclusion

The shown design of a force compensation balance comprises the calculation of the maximum measuring range of 0.5 mN with a resolution of 11 nN for the entire measurement range. The study presents a micro technological workflow design as well as a dynamic model of the designed system. For the system control a PID control approach was modelled. The simulation of the PID controllers indicate limitations that emerge from system components such as the capacitance-to-digital converter and voltage source. Nevertheless, the overall results appear overall promising.

References

1. Rajagopalan, J., Tofangchi, A., Taher, M., Saif, A.: Linear high-resolution BioMEMS force sensors with large measurement range. J. Microelectromech. Syst. **19**(6), 1380–1389 (2010)
2. Koch, S.J., Gayle, E.T., Corwin, A.D., de Boer, M.P.: Micromachined piconewton force sensor for biophysics investigations. Appl. Phys. Lett. **89**, 173901 (2006)
3. Mao, Y.: In vivo nanomechanical imaging of blood-vessel tissues directly in living mammals using atomic force microscopy. Appl. Phys. Lett. **95**, 013704 (2009)
4. Beaussart, A., El-Kirat-Chatel, S., Sullan, R., et al.: Quantifying the forces guiding microbial cell adhesion using single-cell force spectroscopy. Nat. Protoc. **9**, 1049–1055 (2014)
5. Rico, F., Roca-Cusachs, P., Sunyer, R., Farré, R., Navajas, D.: Cell dynamic adhesion and elastic properties probed with cylindrical atomic force microscopy cantilever tips. J. Mol. Recognit. **20**(6), 466–495 (2007)
6. Chen, P., Zhao, Y., Li, Y.: Design, simulation and fabrication of a micromachined cantilever-based flow sensor. In: The 8th Annual IEEE International Conference on Nano/Micro Engineered and Molecular Systems, Suzhou, pp. 681–684 (2013)
7. Nwokeoji A.O., Kumar S., Kilby P.M., Portwood D. E., Hobbs J. K., Dickman M.J.: Analysis of long dsRNA produced in vitro and in vivo using atomic force microscopy in conjunction with ion-pair reverse-phase. In: HPLC, vol. 144, p. 4985 (2019)
8. Diethold, C., Hilbrunner, F.: Force measurement of low forces in combination with high dead loads by the use of electromagnetic force compensation. Measure. Sci. Technol. **23**(074017), 7 (2012)
9. Yamakawa, Y., Yamazaki, T., Tamura, J., Tanaka O.: Dynamic behaviors of a checkweigher with electromagnetic force compensation. In: XIX IMEKO World Congress Fundamental and Applied Metrology (2009)
10. Vasilyan, S., Rivero, M., Schleichert, J., Halbedel, B., Fröhlich, T.: High-precision horizontally directed force measurements for high dead loads based on a differential electromagnetic force compensation system. Meas. Sci. Technol. **27**, 045107 (2016)
11. Pratt, J.R., Kramar, J.A.: Si realization of small forces using an electrostatic force balance. In: XVIII IMEKO World Congress Metrology for a Sustainable Development (2006)
12. Shaw, G.A.: Gordon a Milligram mass metrology using an electrostatic force balance. Metrologia **53**, A86 (2016)
13. Schlaak, H.F., Arndt, F., Steckenborn, A., Gevatter, H.J., Kiesewetter, L., Grethen, H.: Micromechanical capacitive acceleration sensor with force compensation. In: Reichl, H. (ed.) Micro System Technologies 90. Springer, Heidelberg (1990)
14. Kraft, M., Lewis, C.P., Hesketh, T.G.: Control system design study for a micromachined accelerometer. In: IFAC New Trends in Design of Control Systems, Smolenice, Slovak Republic (1997)
15. Mertz, J., Marti, O., Mlynek, J.: Regulation of a microcantilever response by force feedback. Appl. Phys. Lett. **62**, 2344 (1993)
16. Shen, Y., Winder, E., Ning, X., Pomeroy, C.A., Wejinya, U.C.: Closed-loop optimal control-enabled piezoelectric microforce sensorsieee/asme trans. Mechatronics **11**, 420 (2006)
17. Lil, J., Chen, H., Li, Y.: Investigation on surface forces measurement using force- balanced MEMS sensor. In: Conference on Nano/Micro Engineered and Molecular Systems (2006)
18. Coskun, M.B, Moore, S., Moheimani, S.O.R., Neild, A., Alan, T: Zero displacement micro-electromechanical force sensor using feedback control. Appl. Phys. Lett. **104**, 153502 (2014)
19. Fettig, H., Wylde, J., Hubbard, T., Kujath, M.: Simulation, dynamic testing and design of micromachined flexible joints. J. Micromech. Microeng. **11**, 209–216 (2001)

20. Linß, S., Gräser, P., Henning, S., Harfensteller, F., Theska, R., Zentner, L.: Synthesis method for compliant mechanisms of high-precision and large-stroke by use of individually shaped power function flexure hinges. In: Uhl, T. (ed.) Advances in Mechanism and Machine Science, Mechanisms and Machine Science, p. 73 (2019)
21. Henning, S., Linß, S., Zentner, L.: detasFLEX – a computational design tool for the analysis of various notch flexure hinges based on non-linear modeling. Mech. Sci. **9**, 389–404 (2018)
22. Darnieder, M., Pabst, M., Wenig, R., Zentner, L., Theska, R., Fröhlich, T.: Static behavior of weighing cells. J. Sens. Sens. Syst. **7**, 587–600 (2018)
23. Lobontiu, N.: Dynamics of Microelectromechanical Systems. Springer, New York (2007)
24. Bao, M.H.: Handbook of Sensors and Actuators, 2nd edn, vol. 8, Elsevier, Amsterdam (2004)
25. Rogge, N. Weiß, H., Kaiser, I., Amthor, A. Hilbrunner, F., Fröhlich, T.: Design of digital controllers for electromagnetic force compensated balances focused on the disturbance transfer function. In: NCSLI International Workshop and Symposium (2016)
26. Isidori, A.: Nonlinear Control Systems, 3rd edn. Springer, London (1995)

Topology Optimization of Magnetoelectric Sensors Using Euler-Bernoulli Beam Theory

Maximilian Krey(✉) and Hannes Töpfer

Advanced Electromagnetics Group, Technische Universität Ilmenau, Ilmenau, Germany
maximilian.krey@tu-ilmenau.de

Abstract. A studied magnetic field sensor is based on resonant operation of magnetoelectric micro-electro-mechanical systems (MEMS). Subsequently to an applied magnetic field, the micro beam changes the eigenfrequency, due to the magnetostrictive effect. Euler-Bernoulli beam theory can calculate eigenfrequencies of bending vibrations of beams with high accuracy. Implementing more complex beam geometries is challenging, thus often the finite element method (FEM) is used. This paper deals with the modeling of prestressed beams with multilayered structure and discontinuities along the beam length using Euler-Bernoulli beam theory. The arising problems are addressed in detail. As an example, the model is applied to the studied magnetoelectric sensor and shows good accordance to FEM simulations. An optimization algorithm is used to find a sensor geometry that leads to high output signals utilizing the developed model as input for the minimization of a target function.

Keywords: Sensor · Magnetoelectric · Beam theory · Optimization

1 Introduction

Magnetoelectric sensors are capable to detect extremely weak magnetic fields. The magnetoelectric effect of a given material induces an electric polarization $\boldsymbol{P}$ subsequently to an applied magnetic field $\boldsymbol{H}$ [1]. The effect can be enhanced by using a composite of a piezoelectric and a magnetostrictive material. The magnetostrictive effect causes mechanical strain ε_{MS} in a material, when a magnetic field is applied. This is similar to how a piezoelectric material reacts to an applied electric field. In a composite, the strain ε_{MS} is transferred to the piezoelectric material, creating electric polarization. Hence, the detour in the mechanical regime allows to utilize a material with high piezoelectric effect and one with high magnetostrictive effect combined. This enables high magnetoelectric coupling factors and thus sensing of magnetic fields to the pico-tesla range [2] which makes biomedical applications possible [3]. Superconductive Quantum Interference Devices (SQUIDs) sense even weaker magnetic fields but need cryogenic temperatures. The proposed magnetoelectric sensor promises similar sensitivity at room temperature.

Figure 1 shows the proposed sensor, a micro bridge of three layers. The bottom layer A is needed for fabrication, the middle layer B is a piezoelectric material, the top layer C

L. Zentner and S. Strehle (Eds.): MAMM 2020, MMS 96, pp. 115–124, 2021.
https://doi.org/10.1007/978-3-030-61652-6_10

has a gap and is made of a magnetostrictive material. If a magnetic field is applied to the sensor, the arising strain ε_{MS} will not relax like in a singly clamped situation. The strain will rather change the beam's mechanical eigenfrequency like a tuned guitar string. This eigenfrequency change Δf_e can be measured with the electric signal produced by the piezoelectric layer. The signal amplitude increases significantly when the beam vibration is driven in its resonance [4]. Resonant sensors using a frequency-shift based concept are known for high sensitivity and robustness against intensity fluctuations [5, 6].

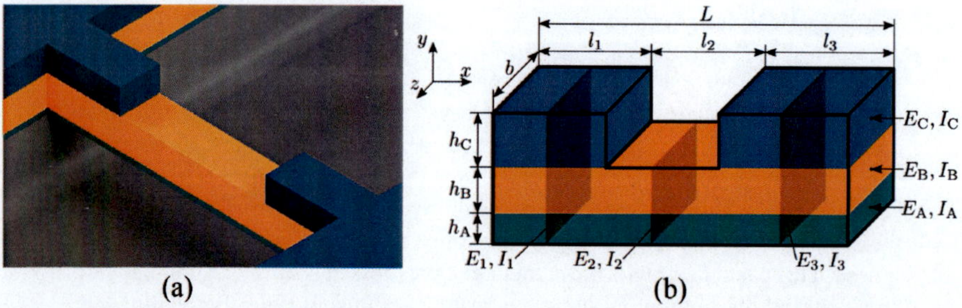

Fig. 1. Rendering of the proposed sensor geometry (a) and model with all parameters (b). E denotes Youngs modulus, I the area moment of inertia, b the beam width, h the layer height, l the section length, and L the beam length.

Full analysis of magnetoelectric sensors is usually complicated because it has to consider simultaneously piezoelectric, magnetostrictive and mechanical phenomena. The paper presents an analytical method focusing on the mechanical part of the sensor, using the approach of continuum mechanics and Euler-Bernoulli beam theory. The presented approach was verified by means of a number of simulations using finite element method (FEM). The applied method was then used to optimize the geometry of the sensor, minimizing a selected target function.

1.1 Euler-Bernoulli Theory for Magnetoelectric Sensors

Derived from the homogeneous wave equation, the Euler-Bernoulli theory describes beam bending vibrations with good accuracy under basic assumption, given in [7]. For example, the beam length should be considerably larger than its height and width. The basic partial differential equation (pde) for the transverse vibration of beams is

$$EI\frac{\partial^4 v}{\partial x^4} + \varrho A\frac{\partial^2 v}{\partial t^2} = 0, \tag{1}$$

where E denotes Young's modulus, I is the area moment of inertia with respect to bending around the z-axis, ϱ and A are the density and the cross-section area of the beam, respectively. The deflection in y-direction is denoted by v, t is time and x is the position along the beam length L. To calculate eigenfrequencies, the separation of variables method is applied in which the deflection $v(x, t)$ is substituted by a product of a time $T(t)$ and the position $Y(x)$ functions. This leads, for $T(t)$, to the ordinary differential equation of a harmonic oscillator. The commonly used analytical solution [8, 9] for solving the position equation is

$$Y(x) = C_1 \cos(\lambda x) + C_2 \sin(\lambda x) + C_3 \sinh(\lambda x) + C_4 \cosh(\lambda x),\ Y, C_{1-4} \in \mathbb{C}, \quad (2)$$

which avoids the imaginary unit by shifting it into constants C_{1-4}. In Eq. (2) λ denotes the eigenvalue. Additionally to Eq. (2), the following boundary conditions are formulated. The micro beam is fabricated by subtractive processes from a surface of the depositioned layers. The passage between the untouched layers and the beam (see Fig. 1a) are the boundaries of the beam model, which are assumed to match a clamping, where the deflection and its gradient (the slope) are zero:

$$v(x_{cl}, t) = 0,\ \frac{\partial v}{\partial x}(x_{cl}, t) = 0. \quad (3)$$

Usually, a clamping is at the beginning or end of a beam, such as $x_{cl} = 0, L$. Since the studied sensor has a gap in the top layer, a discontinuity arises at beginning and end of the gap as depicted in Fig. 1. The two discontinuities create three sections in the beam ($i = 1, 2, 3$), which leads to separate position equations $Y_i(x)$. However, they are still connected with transition conditions, formulated at the positions of discontinuity x_d:

$$v_i(x_d, t) = v_{i+1}(x_d, t),\ \frac{\partial v_i}{\partial x}(x_d, t) = \frac{\partial v_{i+1}}{\partial x}(x_d, t). \quad (4)$$

Equation (4) reflects, that no break appears at x_d in the deflection v_i. Hence v_i is continuous in $0 \le x \le L$. Here, three sections ($D = 3$) are discussed, but a beam with $D \in \mathbb{N}$ sections can be modelled as well. The derivation of boundary and transition conditions for the sensor model is shown in the next section.

The neutral layer of the vibration is shifted from the middle, in a single-material beam, to a position depending on shape and Young's modulus of each layer, in a multilayer beam. This impacts the vibrational behavior of the system, specifically, the parameters E, I, and ϱ in Eq. (1). Because the neutral layer is not in the middle, the parallel axis theorem (Steiner's theorem) applies for the second moment of inertia I for the layers A, B, C. The bending stiffness of beam section 1 and 3 are given as

$$E_1 I_1 = E_3 I_3 = E_{\mathrm{A}}(I_{\mathrm{A}} + y_{\mathrm{A}}^2 A_{\mathrm{A}}) + E_{\mathrm{B}}(I_{\mathrm{B}} + y_{\mathrm{B}}^2 A_{\mathrm{B}}) + E_{\mathrm{C}}(I_{\mathrm{C}} + y_{\mathrm{C}}^2 A_{\mathrm{C}}), \quad (5)$$

where the parameters $y_{\mathrm{A,B,C}}$ denote the distance, from the cross-section centroid of the layer to the neutral layer in y-direction. The y-position of the neutral layer y^* of a beam section with $H \in \mathbb{N}$ layers is defined as

$$y^* = \frac{\sum_{j=1}^{H} E_j A_j + 2 \sum_{j=2}^{H} (E_j A_j \sum_{m=1}^{j-1} h_m)}{2 \sum_{j}^{H} E_j A_j}. \quad (6)$$

Equations (5) and (6) are obtained from [10]. Please note, only in Eq. (6) the layers are numbered, in the remaining paper they have letters as subscript. The Product of $y^2 A$ in Eq. (5) yields the Steiner part to the inertia $I_{\mathrm{A,B,C}}$. The bending stiffness $E_2 I_2$ of section 2 lacks the summand with subscript C in Eq. (5), since layer C is missing in this section (see Fig. 1). After calculating the resulting Inertia I_2, respecting the neutral layer position, a second Steiner part needs to be added to I_2, to account for the off-center placement of this section. The distance of the neutral layers of sections 1 and 2 (as well as sections 2 and 3) y_{s2} is used to include this attribute as $y_{s2}^2 A_2$. The density of a section is calculated as the sum of all section layer densities weighted by their layer's cross-section.

2 Modified Magnetoelectric Sensor Model

Due to the fabrication process, the beam is prestressed. The micro beams described in [11] respect prestress by expanding Eq. (1) with a summand containing an axial force N:

$$EI\frac{\partial^4 v}{\partial x^4} + \varrho A\frac{\partial^2 v}{\partial t^2} - N\frac{\partial^2 v}{\partial x^2} = 0. \tag{7}$$

Adding the N-term, still allows the separation as shown for Eq. (1) and leads to the following formula for the eigenvalue

$$\lambda_{\alpha/\beta} = \pm\sqrt{\frac{p^2}{2k^4} + \sqrt{\frac{p^4}{4k^8} + \frac{\omega^2}{k^4}}}, \tag{8}$$

where k and p are defined as follows

$$k^4 = \frac{EI}{\varrho A}, \qquad p^2 = \frac{N}{\varrho A}. \tag{9}$$

Equation (8) defines the correlation between eigenvalue und and angular eigenfrequency and is called the dispersion relation. The position Eq. (2) is described now by

$$Y(x) = C_1\cos(\lambda_\alpha x) + C_2\sin(\lambda_\alpha x) + C_3\sinh\big(\lambda_\beta x\big) + C_4\cosh\big(\lambda_\beta x\big). \tag{10}$$

As stated before, the beam is split into three sections, connected via the transition conditions. The eigenvalues $\lambda_{\alpha/\beta,i}$ depend on the material properties (see Eqs. (8), (9)) which are different in each section. However, as derived from the first transition condition (4), the angular eigenfrequencies ω_i are the same in each section, as observed in the experiment.

Rearranging the dispersion relation (8) reveals

$$\omega^2 = \lambda_{\alpha/\beta,i}^2\left(k_i^4\,\lambda_{\alpha/\beta,i}^2 \pm p_i^2\right), \tag{11}$$

respectively for trigonometric (α) and hyperbolic (β) part of Eq. (10). Using

$$\omega_i^2 = \omega_{i+1}^2, \tag{12}$$

the eigenvalue of every section i is derived, dependent on $\lambda_{\alpha,1}$ and $\lambda_{\beta,1}$.

Every position equation $Y_i(x)$ has four unknown variables, thus 12 conditions are needed to solve the equation for three sections. Four of them are the boundary conditions (3) for $x_{cl} = 0$ and $x_{cl} = L = l_1 + l_2 + l_3$.

The remaining eight are transition conditions, four for each discontinuity. The geometric transition conditions are stated in Eq. (4), which account for two conditions per discontinuity. The dynamic transition conditions are derived by applying the laws of conservation of momentum and angular momentum to forces and moments during bending at the discontinuities as shown in Fig. 2.

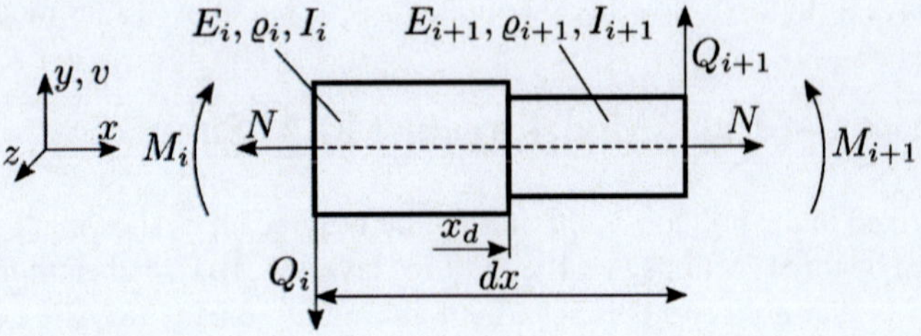

Fig. 2. Shear Forces Q, axial Force N and Moments M at a discontinuity of a prestressed beam during bending around z-axis.

Using the assumptions of classical beam theory as stated in [12]

$$Q = -\frac{\partial M}{\partial x}, \; M = EI\frac{\partial^2 v}{\partial x^2}, \tag{13}$$

the following conditions are derived for both discontinuities

$$-E_i I_i \frac{\partial^3 v_i}{\partial x^3} + E_{i+1} I_{i+1} \frac{\partial^3 v_{i+1}}{\partial x^3} = 0 \tag{14}$$

$$E_i I_i \frac{\partial^2 v_i}{\partial x^2} - E_{i+1} I_{i+1} \frac{\partial^2 v_{i+1}}{\partial x^2} = 0. \tag{15}$$

Please note that the axial force N is not mentioned in the derived conditions and its effect on the vibration is only due to the dispersion relation (11). Figure 2 shows a single layer beam, thus no discontinuity in the neutral layer arises. The resulting moments induced by the axial force N are neglected.

All above mentioned conditions give a system of algebraic equations of the form

$$\boldsymbol{G}\vec{c} = \vec{0} \tag{16}$$

where $\boldsymbol{G}$ denotes the system matrix and $\underline{\vec{c}}$ a vector of unknown variables

$$\vec{c} = \left(C_{1,i}, C_{2,i}, C_{3,i}, C_{4,i}, C_{1,i+1}, .., C_{4,D}\right). \tag{17}$$

The eigenvalue can be found by setting the determinant of $\boldsymbol{G}$ to zero:

$$\det\left(\boldsymbol{G}\left(\lambda_{\alpha,1}, \lambda_{\beta,1}\right)\right) = 0. \tag{18}$$

The numerical search for roots of (18) would have to find $\lambda_{\alpha,1}$ and $\lambda_{\beta,1}$ for one eigenfrequency. A univariate search is created by inserting Eq. (8) in $\det(\boldsymbol{G})$ and finding the angular eigenfrequency ω directly, which reduces computational effort.

Due to the exponential growth of the hyperbolic functions, problems regarding numerical precision arise while handling big numbers. Within the Python programming language [13], the library mpmath [14] allows arbitrary precision, thus this library is used for the calculation of the elements of the system matrix and its determinant.

Please note that matrix $\boldsymbol{G}$ comprises all geometry and material parameters, hence the model can be used to calculate eigenfrequencies of arbitrary beams with discontinuities and multiple layers. In the following, the model is referred to as "Euler model".

3 Comparison of Euler Model Against FEM Simulation

The model derived in the previous section is applied to a beam as shown in Fig. 1, which is distinguished by three sections $i = 1, 2, 3$, three layers A, B, C, and clampings on both sides. Layer A is platinum and B is scandium aluminium nitride (ScAlN), density and Young's modulus are obtained from [15]. The top layer C is made from a magnetostrictive material. Preferably, a material with a high magnetostrictive constant, such as Co/Fe multilayers [16] or Terfenol-D [1], will be applied. In the current experiments, Nickel is selected.

FEM simulations of the beam are performed within Comsol Multiphysics® framework [17]. The pretension is applied by substituting one clamping with a rigid body, with disabled rotational degrees of freedom. A translational movement is only allowed in x-direction (see Fig. 1). Furthermore, the rigid body is fixed to the beam layers and a force N is applied to the rigid body in x-direction. Bending is only allowed in y-direction to inhibit calculation of torsional or out of plane modes, which are out of the scope of this paper.

The first three eigenfrequencies calculated from FEM simulation f_{sim} and the Euler model f_{em} are presented in Table 1. Very small deviations $\delta f_{\%}$ are observed, especially for the first eigenfrequency.

Table 1. Comparison between the first three eigenfrequencies of the simulation and the Euler model.

Order	f_{sim} in Hz	f_{em} in Hz	$\delta f_{\%}$ in %
1	84.978	85.031	-0.062
2	215.657	223.270	-3.530
3	438.203	445.895	-1.755

The behaviour of FEM and Euler-model for a wide range of axial force N is showcased in Fig. 3. The calculations are in good accordance for low forces, they start to deviate around 1000N and reach a state of almost parallel trend for the highest calculated forces. Further investigations are necessary to determine the source of the deviation for high pretensions.

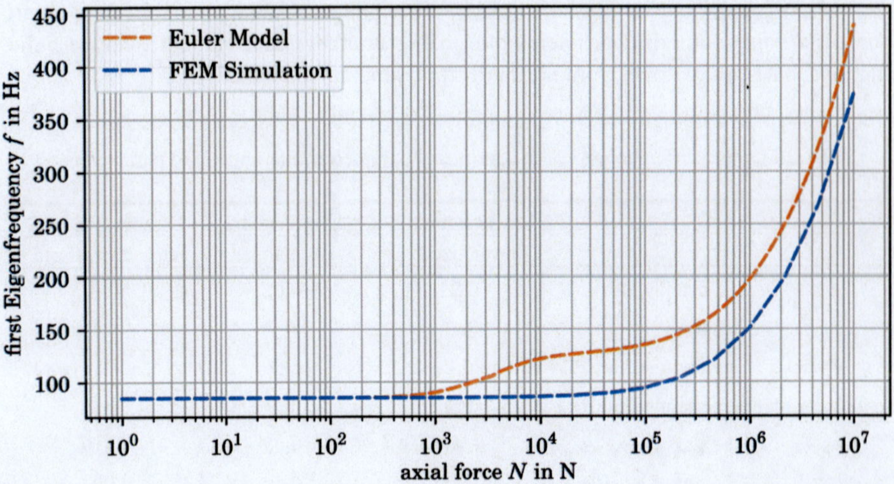

Fig. 3. First eigenfrequency for various axial forces as calculated by FEM simulation and Euler-model.

The beam's geometry is defined by seven parameters: width b, length of all three sections l_i. and height of the three layers $h_{A,B,C}$. The geometry that leads to the frequencies in Table 1 and Fig. 3 is chosen to be in the middle of the bounds stated in Table 2. A beam with this geometry is referred to as standard beam. The bounds are used for a global search in the next section.

4 Sensor Topology Optimization

The magnetoelectric sensor working principle is based on an eigenfrequency shift. The shift is created by changing the tension inside the beam as the magnetostrictive layer has changed its length. Since all layers are connected and the beam is clamped on both sides, it cannot extend or relax but will bend. Buckling is not expected, because the elongation is estimated to be small and the pretension is tensile.

To model the added tension from the magnetostrictive layer and its effect, the eigenfrequencies are calculated twice, first with the force N and second with a force $N + \Delta N$. The frequency shift in percent is defined as:

$$\Delta f_e = \frac{|f_e(N) - f_e(N + \Delta N)|}{f_e(N)} \cdot 100\%. \tag{19}$$

As a target function for minimization, the inverted Eq. (19) is used. It is evaluated for different geometries and fixed material parameters to find a topology that promotes big eigenfrequency shifts. The multivariate numerical optimization algorithm "differential evolution" [18], as provided by Python SciPy library [19], is used. The free geometry parameters are the width of the beam, the length of all sections and the height of all layers. Table 2 lists the parameters (see Fig. 1) and their lower and upper bounds, which are chosen to ensure beam geometries of reasonable size ratios for fabrication. The bounds are given in meters, but should be considered normalized. Calculations for very small geometries raise numerical problems due to big numbers as argument of hyperbolic functions, but scaling down from normalized parameters is possible.

Table 2. Optimization bounds of the geometry parameters and optimized parameters.

Parameter	Name	Lower bound, normalized	Upper bound, normalized	Optimized geometry, normalized
b	beam width	0.01	0.05	0.01
l_1	length of section 1	0.2	0.7	0.353
l_2	length of section 2	0.2	0.7	0.33
l_3	length of section 3	0.2	0.7	0.7
h_A	height of layer A	0.007	0.02	0.016
h_B	height of layer B	0.007	0.02	0.007
h_C	height of layer C	0.007	0.02	0.007

The standard beam has a frequency shift of $\Delta f_{e,st} = 0.007\%$, with optimized geometry a frequency shift of $\Delta f_{e,st} = 0.049\%$ is reached. The steeper slope of the first

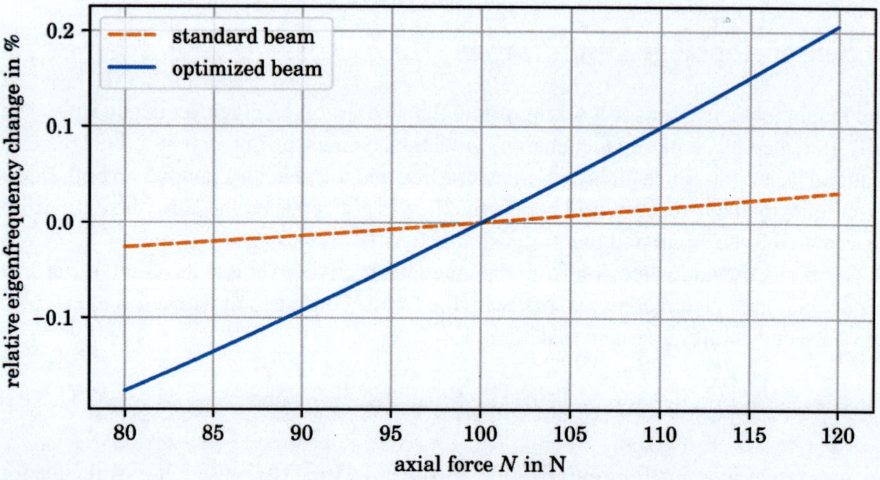

Fig. 4. Change of the first eigenfrequency of standard and optimized beam, as a response to a change of force N.

eigenfrequency of an optimized beam and a standard one, when the force N is varied is shown in Fig. 4. Since the differential evolution algorithm is of stochastic nature, the presented optimized beam geometry might not be the optimal solution. As Table 2 shows, the optimization promotes the smallest possible heights for layers A and B. According to the optimization, an asymmetric beam ($l_1 \neq l_3$) and a small width b is also advantageous for high sensitivities to a change in axial force. Figure 5 shows both standard and optimized beam geometry as rendered images.

(a) (b)

Fig. 5. Qualitative visual comparison of standard (a) and optimized beam (b) geometries (the beam length is scaled down by a factor of 5). Real dimensions are given in Table 2.

5 Conclusion

The Euler-Bernoulli beam theory was expanded to model a magnetoelectric sensor of three layers and three sections with pretension. Every section is assigned a pde describing its vibrational behaviour in dependence on its geometry and material. Resulting Young's moduli, area moment of inertia and densities are calculated, in order to take into account the cross-section and material parameters of the individual layers and the changed neutral layer of the vibration in every section. The pdes are connected via transition conditions. While solving the pdes using boundary and transition conditions, an algebraic system of equations is formed that represents all parameters of the beam. Determined eigenfrequencies of specific beams show good accordance to FEM simulations. The derived method is applied in a topology optimization. The "differential evolution" algorithm calculates the first eigenfrequency of different sensor geometries and evaluates the eigenfrequency shift caused by a change in axial force. The algorithm's target function promotes big eigenfrequency shifts. The derived method aims at developing highly sensitive magnetic field sensors and can be applied to a variety of other problems when model and target function are adapted appropriately.

Acknowledgement. This work was supported by the Free State of Thuringia and the European Social Fund (2017 FGR 0060).

References

1. Nan, C., Bichurin, M., Dong, S., et al.: Multiferroic magnetoelectric composites: historical perspective, status, and future directions. J. Appl. Phys. **103**, 031101 (2008)
2. Bichurin, M., Petrov, V., Petrov, R., et al.: High Sensitivity Magnetometers, Magnetoelectric Magnetometers. Springer, Cham (2017)
3. Hayes, P., Jovičević Klug, M., Toxværd, S., et al.: Converse magnetoelectric composite resonator for sensing small magnetic fields. Sci. Rep. **9**(1), 16355 (2019)
4. Bichurin, M., Fillipov, D., Petrov, V.: Resonance magnetoelectric effects in layered magnetostrictive-piezoelectric composites. Phys. Rev. B Rapid Commun. **68**(13), 132408 (2003)
5. Tilmans, H., Elwenspoek, M., Fluitman, J.: Micro resonant force gauges. Sens. Actuators A Phys. **30**(1–2), 35–53 (1992)
6. Bao, M.H.: Resonant Sensors and Vibratory Gyroscopes. Handbook of Sensors and Actuators, 1st edn. Elsevier Science, Amsterdam (2000)
7. Han, S., Benaroya, H., Wei, T.: Dynamics of transversely vibrating beams using four engineering theories. J. Sound Vib. **255**(5), 935–988 (1999)
8. Irretier, H.: Grundlagen der Schwingungstechnik 2, 1st edn. Vieweg + Teubner, Wiesbaden (2001)
9. Weaver, W., Timoshenko, S., Young, D.: Vibration Problems in Engineering, 5th edn. Wiley India, New Delhi (1990)
10. Bareisis, J.: Stiffness and strength of multilayer beams. J. Compos. Mater. **40**(20), 6 (2006)
11. Bouwstra, S., Geijselaers, B.: On the resonance frequencies of microbridges. In: International Conference on Solid-State Sensors and Actuators 1991, TRANSDUCERS 1991. IEEE, San Francisco (1991)
12. Mang, H., Hofstetter, G.: Festigkeitslehre, 3rd edn. Springer, Wien (2008)
13. Python Software Foundation: Python Language Reference, version 3.7. www.python.org
14. Johansson, F.: mpmath: a Python library for arbitrary-precision floating-point arithmetic (version 0.18) (2013). www.mpmath.org
15. Krey, M., Haehnlein, B., Tonisch, K., et al.: Automated parameter extraction Of ScAlN MEMS devices using an extended euler-bernoulli beam theory. MDPI Sens. **20**(4), 1001 (2020)
16. Honig, H., Hopfeld, M., Schaaf, P.: Preparation and properties of Co/Fe multilayers and Co-Fe alloy films for application in magnetic field sensors. In: Materials Science and Smart Materials, MSSM. Key Engineering Materials, Birmingham (2019)
17. COMSOL Multiphysics® v. 5.4. COMSOL AB, Stockholm, Sweden. www.comsol.com
18. Storn, R., Price, K.: Differential evolution - a simple and efficient heuristic for global optimization over continuous space. J. Glob. Optim. **11**, 341–359 (1997)
19. Virtanen, P., Gommers, R., Oliphant, T., et al.: SciPy 1.0: fundamental algorithms for scientific computing in Python (version 1.3.2). Nat. Methods **17**, 261–272 (2020)

MEMS Acoustical Actuators: Principles, Challenges and Perspectives

Tobias Fritsch, Daniel Beer, Jan Küller, Georg Fischer(✉), Albert Zhykhar, and Matthias Fiedler

Fraunhofer Institute for Digital Media Technology IDMT, Ilmenau, Germany
daniel.beer@idmt.fraunhofer.de

Abstract. 10 years ago, the MEMS technology unexpectedly revolutionized the global microphone market. Will the MEMS technology have similar impact on the loudspeaker market? This article introduces MEMS loudspeakers. It covers the underlying manufacturing technology as well as technical requirements for the headphone and hearing aid market. It covers a section about artificial ear measurements for MEMS loudspeakers. Moreover, the article underlines the importance to use digital signal processing to increase MEMS performance and sound optimization. A section summarizes six different MEMS loudspeaker approaches from the literature. The descriptions contain technical specifications and the mechanical principle of sound reproduction. Finally, the article considers aspects of acoustics in very small systems. Here, thermoviscous boundary layer effects dominate acoustics.

Keywords: MEMS · Loudspeaker · Headphone · Signal processing · Nonlinear · Thermoviscous · Boundary layer

1 Introduction

Advances in the field of manufacturing technology enable new product solutions. These may consist of completely new approaches, but also of familiar ones, which have not yet been implemented satisfactorily with the new technology.

There is a high demand for miniaturized, high-performance loudspeakers in the field of consumer electronics and medical technology. Typical applications are headphones, hearing aids, hearables, and smart glasses. The loudspeakers used today are still produced using precision engineering processes. Different machines have to combine different components, like magnet, voice coil, membrane, etc. Additionally, the loudspeaker requires electrical connections to further parts like sensors, amplifiers, and other integrated circuits. These components come from different manufacturers and have incompatible manufacturing technologies. Nonetheless, they have to be closely integrated to realize advanced loudspeaker control. The combination of mechanical and electrical features in one device is hence complex. This causes limitations to miniaturization, accuracy, and cost-effective implementation.

L. Zentner and S. Strehle (Eds.): MAMM 2020, MMS 96, pp. 125–136, 2021.
https://doi.org/10.1007/978-3-030-61652-6_11

This changed with MEMS; the combination of miniaturized mechanical and electrical components in a single system offers completely new perspectives for meeting the loudspeaker requirements of the market.

1.1 Micro-Electro-Mechanical Systems

The acronym MEMS stands for **M**icro-**E**lectro-**M**echanical **S**ystem. MEMS include both miniaturized electrical and miniaturized mechanical functional elements. Foundries and original equipment manufacturers (OEMs) produce MEMS with the technologies of the semiconductor industry. These technologies include additive and subtractive processes like vapor deposition and etching. Figure 1a illustrates an example of this manufacturing process. The starting material is usually a circular wafer made of highly pure silicon. Depending on chip size, the wafer can hold thousands of MEMS. This allows for effective parallel manufacturing. After processing, the MEMS will be separated by dicing. Finally, the MEMS are housed. Figure 1b illustrates these steps from a raw silicon wafer to housed MEMS chips.

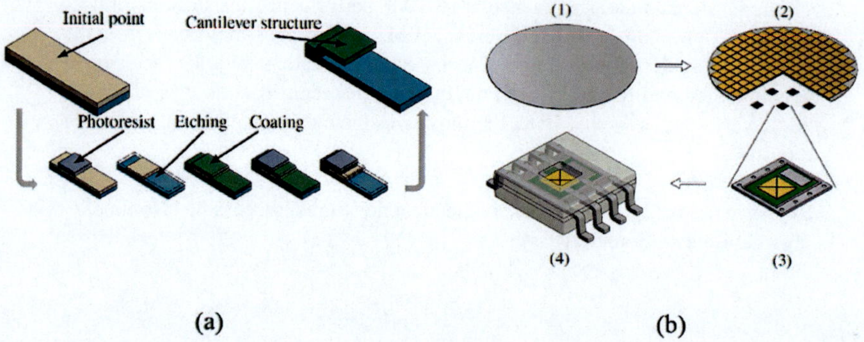

Fig. 1. a) Illustration of surface micromachining process [1]. b) Steps of wafer–based MEMS manufacturing, starting with the raw wafer (1), processing of chips on the wafer (2), separation into single chips (3), and finally the packaging (4) [1].

According to the designation MEMS, it is common to integrate the control circuit in form of an **A**pplication **S**pecific **I**ntegrated **C**ircuit (ASIC) on chip or in package. The package can also contain several other functional units, like sensors, Bluetooth receivers, and amplifiers. This high degree of integration helps to prevent energy losses and decrease the necessary installation space.

1.2 MEMS in Mobile Devices – A Motivation for MEMS Loudspeakers

All modern mobile device are equipped with a variety of MEMS modules: accelerometers, thermometers, antenna tuners, and more. Of special interest are MEMS microphones. They have continually substituted electret microphones in portable audio since 2014 [2]. The MEMS microphone has proven the market-potential of acoustical MEMS. The MEMS microphone market reached a value of over 1,000 Mio. $ in 2016. In this year, the MEMS industry sold over 4,400 Mio. MEMS microphones [2]. One reason for this success is that MEMS technology achieves a significant reduction of the microphone size and price with still better performance and robustness. Additionally—and importantly—MEMS microphones offer enhanced solderability.

However, development started much earlier. First designs for MEMS microphones were already being developed at the beginning of the 1980s [3].

2 MEMS Loudspeaker – Actuators for Human Hearing

The nominal human ear is sensitive to linear and nonlinear distortion, noise, delay and to a certain extent even to phase shifts [4, p. 360]. It has a dynamic range of 120 dB [4, p. 133] and is able to perceive frequencies in a range from 16 Hz up to 20 kHz [4, p. 160]. The sensitivity of the individual ear, however, depends on physical factors and training. Even the anthropometrics of the outer ear, head, and upper torso affect the sound, causing diffraction and reflections [4, p. 264]. In-ear headphones circumvent these effects by emitting sound directly into the ear canal. Without the influence of the outer ear, we perceive sound in an unusual way, which does not feel natural. Headphone manufacturers usually define a target curve to address this problem. The target curve describes the desired frequency response of headphones that reproduces the perceived frequency response of a good set of speakers in an optimized room. Many studies have derived different target curves with varying methods and changing number of test participants [5]. The physiological differences between participants are, however, too strong to derive a generally applicable model of the human ear. Hence, it is not possible to obtain a target curve that is suitable for everyone. Section 2.1 provides technical requirements for in-ear loudspeakers.

To evaluate their headphones, manufactures use artificial ears. Section 2.2 introduces this kind of measurement. Knowledge of the measurement principle is crucial to the understanding of the following results of measurements and simulations. Finally, Sect. 2.3 suggests using digital signal processing to improve the performance of MEMS loudspeakers and adapt them to the requirements of individual hearing.

2.1 Loudspeaker Requirements for Earphones and Hearing Aids

Traditionally, earphones and hearing aids are dissimilar products. The use cases differ and the development of these devices has been an independent process. Headphones and earphones rely mainly on electrodynamic speakers with a moving coil. The hearing aid industry utilizes balanced armature speakers. Hearing aids had to rely on batteries from the beginning, while the widespread introduction of wireless earphones happened recently. Earphones where required to deliver a broad acoustic spectrum with moderate **s**ound **p**ressure **l**evels (SPL), while hearing aids need to deliver very high levels in the limited frequency range of human speech.

This distinction between earphones and hearing aids is beginning to fade, however—especially in the areas of light to medium hearing losses. The customers expect perception-enhancing functionality from their smart devices like hearables. New deregulation laws in the US further accelerate this process [19]. Because of this merging of industries, a new type of loudspeaker will be required, which is able to fulfill the requirements of both applications. Table 1 shows these minimum requirements. It was derived from interviews with experts from both headphone and hearing aid industry [7, 8].

Table 1. Loudspeaker requirements for earphone, hearable and hearing aid devices [7, 8].

Parameter	Earphone/Hearable	Hearing aid
Sound pressure level	≥100 dB	≥120 dB
Total harmonic distortion	<1 %	<5 %
Bandwidth (−20 dB)	20–20,000 Hz	100–6,000 Hz
Sensitivity	105 dB/1 mW	105 dB/1 mW
Operating voltage	3.6 V	3.6 V
Battery lifetime	4–10 h	>14 h
Size [mm]	(Ø, h) 7–14, 3–8	(LWH) 9 × 6 × 3
Weight	≤2 g	≤1 g
Price	<3 USD	<5 USD

2.2 Characterizing MEMS Loudspeakers

This section considers some aspects to enable the accurate measurement of MEMS loudspeakers and a better understanding of the results thereof. The section is not specific to MEMS but of importance to everyone interested in MEMS loudspeakers and not familiar with artificial ear measurements. Possible problems with electrical measurements are briefly described.

Acoustical Measurements. MEMS loudspeakers are small and target the in-ear or near-ear market. All kinds of in-ear headphones utilize the pressure chamber between loudspeaker membrane and eardrum. This pressure chamber will lead to eardrum movements, which are directly proportional to membrane movement. This works up to a certain frequency, which depends mainly on the air volume and possible leakages [9]. Small loudspeakers need the pressure chamber to reproduce high pressure levels at low frequencies. In free-field conditions, low frequencies from small speakers are hardly detectable, even from short distances. This motivates to include the pressure chamber effect into the measurement principle. Couplers enable this; they provide a defined volume between speaker and microphone.

Artificial ears are expanded couplers, which also model the transfer function and acoustic impedance of the inner ear [10]. There are different specifications. The artificial ear described in IEC 60318–4 [11] is often the right choice. This artificial ear models an average human ear in the frequency range of 100–10,000 Hz. There is also a high-frequency version, valid up to 20,000 Hz.

When examining headphone measurements it is important to consider the transfer impedance of the artificial ear. Figure 2 illustrates the typical transfer impedance of an

IEC 60318–4 artificial ear. The high resonance peak above 10 kHz will affect SPL measurements. Different insertion depths of the earphone into the ear canal will shift this resonance peak. Additional volumes or leakages between artificial ear and speaker can introduce significant deviations in the measurement results. To achieve reproducible results, one should always average at least five measurements with individual re-insertions.

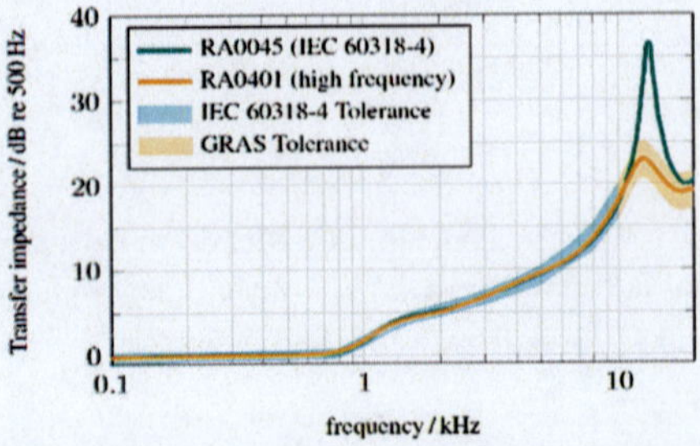

Fig. 2. Transfer impedance of typical IEC 60318–4 artificial ear (GRAS RA0045) in comparison to a new high-frequency model (GRAS RA0401). Colored areas show the corresponding tolerances [12].

Electrical Measurements. A huge challenge arises when performing impedance analysis, power rating characterization, or other electrical measurements on MEMS loudspeakers. The electrical impedance of a MEMS loudspeaker can be higher than 100 MΩ [13], which is higher than the impedance of normal oscilloscopes (1–10 MΩ). Highly specialized measurement hardware in the nanoamperes range is required. For such low currents, outside effects can influence the measurement easily and will introduce relative errors of high magnitude.

2.3 Controlling MEMS with DSP

Digital **S**ignal **P**rocessing (DSP) can be used to compensate linear and nonlinear signal distortion. Nonlinear distortion is caused by the driving mechanism and/or high membrane deflections. High membrane deflections are needed to push MEMS loudspeakers to their limits, which in turn is necessary to obtain enough sound pressure from the usually small membrane surfaces. The listener will perceive even small amounts of nonlinear distortion as unpleasant. For this reason, every MEMS loudspeaker will need DSP when used to its full potential. Figure 3b illustrates the reduction of **T**otal **H**armonic **D**istortion (THD) using feedforward control. To address different hearing habits and to ensure good sound quality, we apply equalization to the transfer function of the Fraunhofer ISIT MEMS loudspeaker. Figure 3a shows the original and the resulting frequency response.

Due to the high quality factors of the resonances in the small Fraunhofer ISIT MEMS loudspeaker structures, slight shifts of the resonance due to aging effects or changing acoustic impedance may cause the designed filters to be useless or even self-defeating.

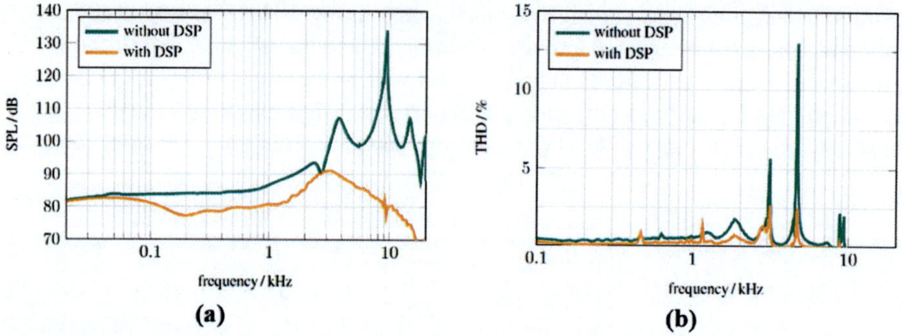

Fig. 3. a) Measured SPL at 1 V_{AC} (peak–peak) of a MEMS in-ear-headphone with and without processing. The input signal is filtered (FIR-EQ) to match the earphone response with a target curve. b) THD curve under condition of Fig. 3a. The measurements were performed in a GRAS RA0401 artificial ear [14].

This leads to the need for adaptive control strategies to update the filters according to the system state. For this, sensors need to be included in the final system design. These can be other MEMS-based devices like microphones, accelerometers, photodiodes, etc. Another way to sense the state of the system lies in measuring the voltage and current at the input of the loudspeaker. Depending on the MEMS design, this may be difficult, as discussed in Sect. 2.2. Depending on the chosen sensor, a suitable model of the system has to be developed that mimics the system behavior and/or underlying physics. For this, the MEMS loudspeaker needs to be characterized by thorough measurements and simulations.

DSP also offers features like protection of the MEMS loudspeaker against over excursion, overheating, and electric breakdown. Both may play a crucial role, depending on the design and underlying driving principle of the loudspeaker.

3 Approaches to MEMS Loudspeakers

Starting from 1994, various research institutes and companies have published new scientific articles and patents about MEMS loudspeakers. Some concentrate on ultrasonic sound radiation. A loudspeaker for hearable sound needs to move larger volumes of air and has to handle a larger relative bandwidth. Therefore, loudspeakers for audio applications are a greater challenge and will be the subject of this discussion. The following section introduces six significant approaches to MEMS loudspeakers. These six MEMS loudspeakers are only a subset of the known literature but represent the most fundamental driving mechanisms. Figure 4 shows illustrations of these approaches.

3.1 Initial MEMS Loudspeaker Designs

Piezoelectric MEMS Loudspeaker (1994). Lee et al. published a piezoelectric cantilever MEMS loudspeaker. The speaker has a 2 mm × 2 mm cantilever-membrane, which has a fixed bearing on one side. The other sides are separated from the surrounding structure by a 10 μm gap. On top of this cantilever, near the fixed side, is a

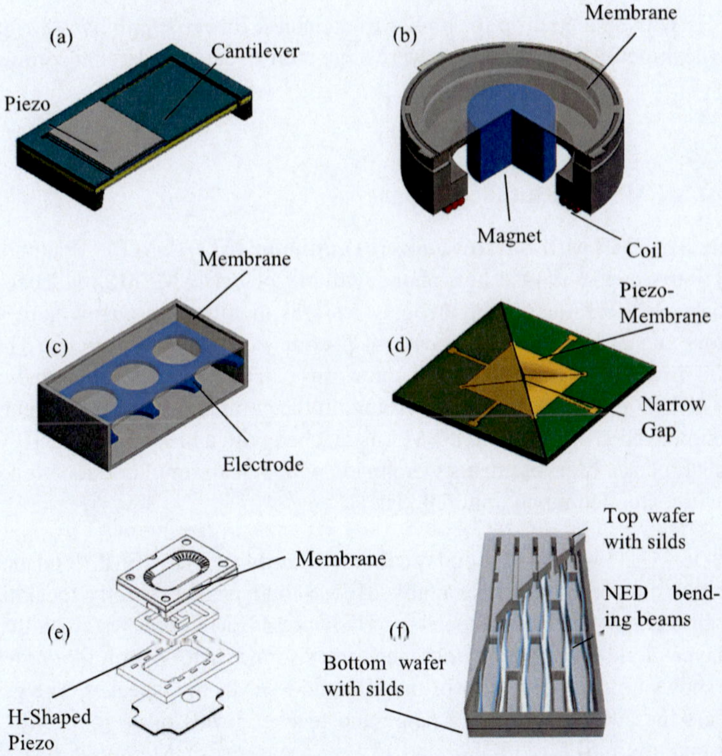

Fig. 4. Different MEMS approaches a) Lee et al. [15]. b) Haradine et al. [16] c) Loeb et al. [17] d) Fraunhofer ISIT/IDMT [14] e) USound [1] f) Fraunhofer IPMS/Arioso [13].

piezoelectric actuator. It causes oscillation of the cantilever-membrane when voltage is applied. A sectional drawing can be found in Fig. 4a. Lee et al. were able to generate air-borne sound covering the audible frequency range. The sound pressure level reaches 100 dB at 4.8 kHz with amplitude of 12 V_{AC} (peak-peak). The frequency response, however, shows a strong drop of approximately 40 dB around 2.5 kHz. Measurements were done with a 2 cm^3 coupler [15].

Electrodynamic MEMS Loudspeaker (1997). Haradine et al. introduced a MEMS loudspeaker with electrodynamic drive system. They attached a permanent magnet to a circular membrane with a diameter of 2.5 mm. Membrane and magnet move together, while the voice coil stays fixed in the surrounding substrate. A sectional drawing can be found in Fig. 4b. Maximum SPL is around 35–50 dB from 100–1,000 Hz with a narrow peak up to nearly 100 dB at 3.2 kHz. Measurements where done with a 2 cm^3 coupler [16].

Electrostatic MEMS Loudspeaker (1999). Loeb et al. introduced one of the first electrostatic driven MEMS loudspeakers for sound reproduction in audio frequency range. A square membrane with an edge length of about 1.4 mm interacts with a stator electrode behind. A sectional drawing can be found in Fig. 4c. They achieve 70–80 dB from

20–4,000 Hz with a steep drop for higher frequencies. Power supply was 80 V_{DC} with 28.6 V_{AC} (peak-peak). Measurements were done with an ear-simulator according to IEC 60318–4 [17].

3.2 Latest MEMS Loudspeaker Designs

Piezoelectric MEMS with Narrow Gaps (Fraunhofer ISIT/IDMT). Fraunhofer ISIT and IDMT introduced an in-ear headphone with piezoelectric MEMS loudspeaker. The four piezoelectric actuators simultaneously work as membrane. Narrow gaps separate the actuators and allow free movement but prevent acoustic short-circuit. The active chip area is 4 mm × 4 mm. Figure 4d shows two deflected actuators and two in rest position. The headphone can reproduce sound in the complete audible frequency range with a maximum SPL of 110 dB. This SPL is reached with a bias voltage of 10 V_{DC} and 20 V_{AC} (peak-peak). Measurements were made with an ear-simulator according to IEC 60318–4 with extended range up to 20 kHz [12].

Piezoelectric MEMS-Hybrid Loudspeaker (USound). The MEMS loudspeaker by USound is the first commercially available MEMS loudspeaker. Strictly speaking, it is a MEMS-hybrid consisting of a piezoelectric H-shaped MEMS drive system with a glued-on multi-layer membrane. The membrane creates sounds with piston like movements. Figure 4e shows an exploded view of the USound MEMS loudspeaker. The generated sound covers the audible frequency range and reaches levels of up to 110 dB. Power supply for maximum SPL is 15 V_{DC} and 15 V_{AC} (peak-peak). Measured in an artificial ear according to IEC 60318–4 [1].

Nanoscopic Electrostatic Drive (NED) MEMS Loudspeaker (Fraunhofer IPMS). Fraunhofer IPMS designed their MEMS loudspeaker to use the volume of the chip for sound reproduction, not only the surface. This is achieved by using bending beams inside the chip. These electrostatically driven beams push the air through the MEMS loudspeaker and create hearable sound. Figure 4f illustrates the beams of only 75 μm height. Also visible are slits in bottom plate and lid. The loudspeaker produces a maximum SPL of 100 dB for a bias voltage of 40 V_{DC} and a driving voltage of 10 V_{AC} (peak-peak) with a chip surface of 6.2 mm × 2.4 mm. Measurements were conducted with an artificial ear according to IEC 60318–4 [13].

4 New Design Challenges

By miniaturization down to the micrometer level, additional physical phenomena have to be taken into account that are negligible for conventional actuators. This refers to thermoviscous losses, which affect the propagation of sound waves and fluid flow in narrow structures. Tangential forces occur at the transition between fluid and the wall surface of a solid. Such forces result in friction, which extracts energy from the wave field or fluid flow, respectively, and converts it into heat [18, p. 298].

4.1 Thermoviscous Acoustics

In acoustics, there is a distinction between viscous and thermal boundary layer. The viscosity of a fluid, like air, in sound channels causes a decrease in sound velocity towards the wall surface. At the wall surface of the solid no more wall parallel oscillation takes place (Fig. 5). This transition is called the viscous boundary layer. Simultaneously, heat is transported between the medium and the wall surface, as the thermal conductivity of air is several orders of magnitude smaller than that of solids. This transition indicates the thermal boundary layer. Both boundary layer thicknesses are frequency dependent [18, p. 323]. Figure 6a shows the viscous δ_{visc} and thermal δ_{therm} boundary layer thickness in air compared to the acoustic wavelength λ. The theoretical principles of thermoviscous losses are discussed in more detail in [19].

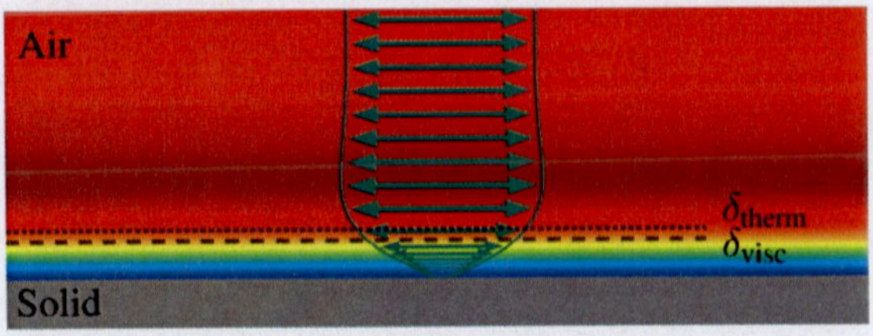

Fig. 5. Velocity profile of air particles due to sound propagation at a solid surface with viscous δ_{visc} and thermal δ_{therm} boundary layer thicknesses.

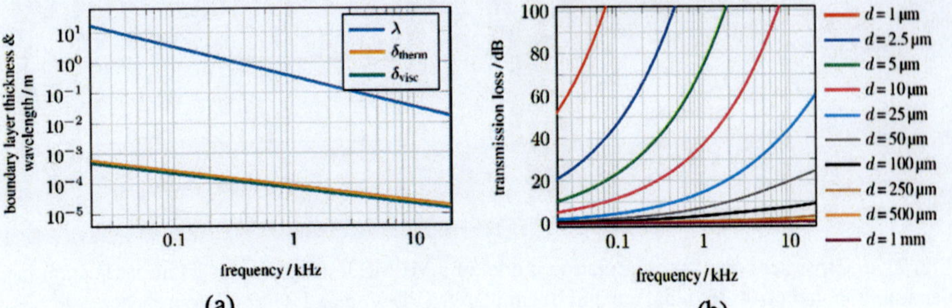

Fig. 6. a) Viscous δ_{visc} and thermal δ_{therm} boundary layer thicknesses, and acoustic wavelength λ in air as a function of frequency. b) Simulated attenuation of the SPL in a circular channel after 1 cm depending on the diameter d and the frequency

4.2 Damping Due to Thermoviscous Losses

Numerical computer simulations performed with COMSOL Multiphysics® determine the geometric dimensions at which thermoviscous losses cause an attenuation of the SPL. An infinitely long circular channel is the basis of the simulation model. The model considers the linearized Navier-Stokes equations, with valid no-slip and isothermal boundary conditions [20, Ch. 2, pp. 19–42]. A plane wave with a sound pressure amplitude of p_{in}

= 1 Pa at the beginning of the channel excites the included air to oscillate. Two measurement points at a distance of 1 cm are used to determine the attenuation of the SPL for circular channels of different diameters d as shown in Fig. 6b. The attenuation for channels with diameters $d > 100$ μm is below 10 dB/cm. For decreasing diameters, the attenuation increases rapidly. At $d < 25$ μm the viscous and the thermal boundary layers reach the dimensions of the channel for the entire audible spectrum. Therefore, friction appears over the whole channel.

4.3 Thermoviscous Losses in Practice

A simulation model of the Fraunhofer ISIT/IDMT piezoelectric MEMS in-ear headphone from [14] is shown in Fig. 7a. It demonstrates the importance of thermoviscous losses and the resulting acoustic impedances. In this case, thermoviscous acoustics were applied to the relevant regions. The MEMS loudspeaker is encapsulated in an in-ear headphone enclosure that plugs into an artificial ear and includes a back volume and sound channel. The sound channel radiates into the free field. For comparison, simulations with and without thermoviscous losses were performed. The MEMS loudspeaker was driven at 1 V_{AC} (peak-peak) and 1 V_{DC}. Additionally, a measurement validates the results in Fig. 7b.

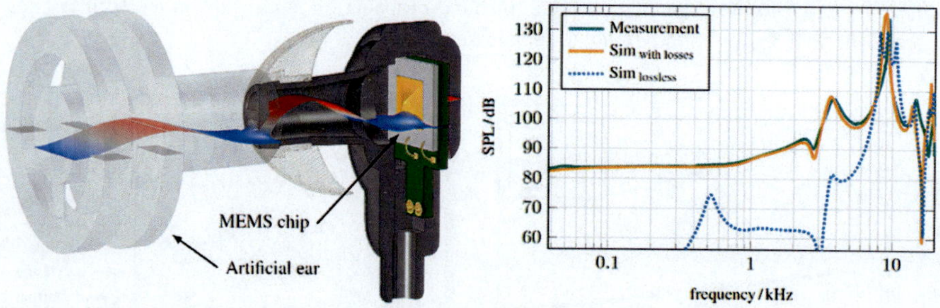

(a) (b)

Fig. 7. a) Cross section of the simulation model of a MEMS loudspeaker as an in-ear headphone version mounted on an artificial ear and b) simulated SPL with and without thermoviscous losses, including measured SPL [14] of the same headphone.

When simulating with thermoviscous losses, the calculated acoustic impedances, damped resonances, and cancellations in the frequency response are in good agreement with the measurement. In the lossless simulation, the acoustic impedances inside narrow geometries are no longer modeled correctly. Against expectation, this results in a drop of lower frequencies, additional resonances, and cancellations. This can be explained by two factors. First, the slits of the membrane are no longer acoustically sealed, an acoustic short-circuit is the consequence. Second, the output impedance of the channel at the back of the headphone housing is much smaller than the impedance of the artificial ear: Sound energy is lost into the free field. This example shows that thermoviscous losses are not negligible in the design process of miniaturized acoustic actuators and sound guides with small dimensions.

4.4 Utilization of Thermoviscous Losses

Thermoviscous losses are especially noticeable for resonances and cancellations. The reason is that the sound velocity reaches its maximum within small structures at resonances. The smaller such structures, the more the sound velocity is slowed down and thus the more the resonances will be attenuated. On the one hand, this leads to performance losses. On the other hand, this can be utilized in the design process to minimize disturbing sound effects. Today's manufacturing methods, such as 3D printing or injection molding, allow the production of small topologies. Consequently, there is a better chance to achieve target SPL curves by developing well-engineered acoustic concepts. Nevertheless, these systems need a suitable signal processing to utilize their full potential.

5 Conclusion

The latest MEMS loudspeaker designs, as presented in Sect. 3.2, show a high potential of MEMS technology for speaker manufacturing. But even these designs do not fulfill all the requirements of the headphone industry. The biggest advantage of the MEMS loudspeaker is also its biggest challenge: the miniaturization. It forces engineers to drive the loudspeakers to their physical limits and thermoviscous losses are no longer negligible. Luckily, numerical simulations are a very powerful tool for the design and optimization of MEMS loudspeakers, including thermoviscous effects. Here, the choice of the right physical model is most important.

Driving the MEMS to its physical limits will cause thermal, mechanical, and electrical stress and lead to noticeable nonlinear behavior. Nonlinearities can be handled by an adaptive loudspeaker control strategy. The necessary sensor signal can be derived from an integrated sensor. Luckily, MEMS offer the possibility to integrate signal processing circuitry and sensors directly on chip.

The validation of MEMS speakers can be done via measurements. The measurement results differ depending on the used setup. To compare the results, measurements have to be performed under the same conditions. That includes the used measurement hardware, the same integration of the device under test into the setup and the same environmental parameters.

References

1. Beer, D., Rusconi, A., Stoppel, F., Ehrig, L.: MEMS-Lautsprecher – ein Paradigmenwechsel. DEGA Akustik J. **2** 4–39 (2020)
2. Acoustic MEMS and Audio Solutions – From Technologies to Markets, Yole Developpement, May 2017
3. Royer, M.: Piezoelectric pressure sensor, U. S. Patent Application Filed, 30 March 1982
4. Moore, B.: Introduction to the Psychology of Hearing, 6th edn. Brill, Leiden (2013)
5. Olive, S., Welti, T., Khonsaripour, O.: Segmentation of listeners based on their preferred headphone sound quality profiles. AES Convention, Dublin, vol. 146 (2019)
6. Letter of the FDA to Hearing Aid Manufacturers. https://www.fda.gov/media/114844/download. Accessed 10 June 2020

7. Richter, H., Audifon GmbH & Co. KG, Kölleda/Germany, Private Communication with Daniel Beer, Fraunhofer IDMT, Ilmenau/Germany, 26 November 2019
8. Koch, T.: Private Communication with Daniel Beer, Fraunhofer IDMT, Ilmenau/Germany, 27 November 2019
9. Poldy, C.: Headphones (chap. 14). In: Browick, J. (ed.) Loudspeaker and Headphone Handbook, 3 edn., pp. 585–692. Focal Press, Oxford (2001)
10. Series P: Terminals and Subjective and Objective Assessment Objective Measuring Apparatus Artificial Ears, Recommendation ITU-T P.57 12.2011
11. Electroacoustics - Simulators of human head and ear - Part 4: Occluded-ear simulator for the measurement of earphones coupled to the ear by means of ear inserts, IEC 60318–4:2010 (2010)
12. GRAS Sound & Vibration, Instruction Manual GRAS 60318–4 Ear Simulators. Recreated with data kindly submitted by GRAS, November 2019
13. Ehrig, L., Kaiser, B., Schenk, H.A.G., Stolz, M., Langa, S., Conrad, H., Schenk, H., Männchen, A., Brocks, T.: Acoustic validation of electrostatic all-silicon MEMS-speakers. In: Audio Engineering Society Conference: 2019 AES International Conference on Headphone Technology, August 2019
14. Männchen, A., Stoppel, F., Brocks, T., Niekiel, F., Beer, D., Wagner, B.: Design and electroacoustic analysis of a piezoelectric MEMS in-ear headphone. In: AES Conference on Headphone Technology (2019)
15. Lee, S.S., Ried, R.P., White, R.M.: Piezoelectric Cantilever Microphone and Microspeaker. Solid-State Sensor and Actuator Workshop Hilton Head, South Carolina (1994)
16. Harradine, M., Birch, T., Stevens, J., Shearwood, C.: A micro-machined loudspeaker for the hearing impaired. In: Proceedings of International Solid State Sensors and Actuators Conference (Transducers 1997), Chicago, vol. 1, pp. 429–432, IEEE (1997)
17. Neumann, J.J., Gabriel, K.J.: CMOS-MEMS membrane for audio-frequency acoustic actuation. Sens. Actuators A: Phys. **95**(2), 175–182 (2002). Papers from the Proceedings of the 14th IEEE International Conference on MEMS Systems
18. Blackstock, D.T.: Fundamentals of Physical Acoustics. Wiley, New York (2000)
19. Küller, J., Zhykhar, A., Beer, D.: Sound Propagation in Microchannels. In: DAGA (2020)
20. Bruus, H.: Theoretical Microfluidics. Oxford University Press, Oxford (2010)

Synthesis of Micro-robotic Appendages Considering Different Performance Properties

Andrija Milojević[1] (✉) and Kenn Oldham[2]

[1] LUT University, 53850 Lappeenranta, Finland
Andrija.Milojevic@lut.fi
[2] University of Michigan, Ann Arbor, MI 48109, USA
oldham@umich.edu

Abstract. This paper presents a synthesis method for microelectromechanical system (MEMS) terrestrial micro-robotic appendages considering their different performance properties. In micro-robotics, actuation is often a challenge. Among candidate actuator types, thin-film lead-zirconate-titanite (PZT) may be adopted due to potential advantages in energy density and voltage and current requirements. Proposed micro-robotic appendages may be comprised of multiple micromachined materials: parylene-C (soft, passive), silicon (rigid, passive) a and thin-film PZT laminates (active). But a synthesis method for such type of micro-robots is lacking. In this paper, the topology optimization synthesis approach is introduced considering placement (type, dimensions, location, and orientation) of both active (thin-film PZT) and passive (parylene-C and Si) multiple materials. The goal of the optimization is to realize micro-robotic appendages solutions, where transport ability and projected walking speed are maximized while needed electrical power is minimized. The individual steps in the synthesis process are presented together with different obtained solutions. Finite element method simulations of the micro-robotic appendage deformation behavior are presented as well.

Keywords: Micro-robot · Appendages · Synthesis · Thin-film PZT · MEMS

1 Introduction

In many proposed robot applications, classical large-scale robots are not suitable and cannot be used. This is especially the case when robots need to access small and narrow places to perform functions such as inspection, delivery, sample collection, navigation of highly unstructured or hazardous areas, realization of targeted drug delivery, or exploration in lunar or space environments. For such kinds of applications, micro-robots have been proposed [1–3]. The use of micro-robots offers many advantages including minimal disturbance to users, small power requirements, portability, disposability, and ability to work in large teams and exploit swarm intelligence. Thus, micro-robotics is a very relevant and ongoing research topic that has received much attention in recent years [1–8]. Among many different types of micro-robots [1, 2], MEMS-based terrestrial micro-robots are of special interest in this paper (Fig. 1a). Such kinds of robots are

L. Zentner and S. Strehle (Eds.): MAMM 2020, MMS 96, pp. 137–148, 2021.
https://doi.org/10.1007/978-3-030-61652-6_12

often produced with MEMS (microelectromechanical system) technologies where robot size usually ranges from few millimeters to 1–2 cm. MEMS micro-robots offer several advantages such as highly integrated compact systems and low-cost manufacturing due to batch fabrication.

But at the micro-scale, there is often a problem of providing actuation to the micro-robots. Different types of actuation technologies have been proposed over time [4–6]. The realization of MEMS micro-robots or more specifically micro-robotic appendages with thin-film PZT as actuators (Fig. 1b) were of special interest in the authors' past work [7, 8]. Thin-film piezoelectric materials (PZT) offer many advantages including high actuation speeds, small power consumption, high work density per unit area, and reasonable efficiency as small capacitive loads [5], relative to other MEMS transducers. Drawbacks are limited payload, small inherent stroke lengths (a small range of motion for the micro-robot), and complex fabrication processes. It has been shown [7] that the use of thin-film PZT's in combination with other materials like polymer (soft) and silicon (Si, rigid) elements can overcome some of the disadvantages of PZT actuators. Moreover, it was shown that robustness and range-of-motion of resulting micro-robotic appendage could be significantly improved.

Thus, this paper examines MEMS terrestrial micro-robots that realize their locomotion via appendages comprised of parylene-C (soft) and Si (rigid) structures as passive elements, and thin-film PZT as active elements (Fig. 1c and d). But this raises the question of how to design such micro-robotic appendages. Typically, micro-robotic appendages are designed based on designer experience, through a trial-and-error approach or taking inspiration from nature (mostly insects) [3, 7, 8]. Realizing micro-robotic appendages in such a way often leads to non-optimized designs where the appendages' range of motion and performance are limited. Thus, the goal of this paper is to develop a synthesis method for the realization of optimized micro-robotic appendages. Here the designer would only need to specify desired initial parameters, and through the algorithmic procedure, the optimal solution would be automatically obtained, where prior design experience is not needed.

As a foundation for the synthesis method, the topology optimization approach is adopted here [9, 10]. Topology optimization has been considered before for the design of MEMS-based structure as shown in many examples [9, 11], but none has dealt with the synthesis of more complicated MEMS devices such as micro-robots. For the synthesis of micro-robotic appendages, placement (type, dimensions, location, orientation) of both active (thin-film PZT) and passive (parylene-C and Si) elements needs to be considered simultaneously in the optimization process. Moreover, here multiple materials need to be considered as well. In contrast, in [11] the actuators are added after the MEMS structure (comprised of passive elements) is obtained with optimization. Topology optimization including multiple materials was presented in [10], but not including the placement of active elements. In [12] the synthesis of microstructures comprised entirely of active elements is presented, but not including placement of passive multiple-material components. Further, synthesis of structures with simultaneous placement of both active and passive elements is considered in [13], but not including multiple materials or considering specific actuator type (like thin-film PZT). Continuum based topology optimization considering only multiple passive materials [11] and both active and passive materials (with limited placement variables) [14] has been realized before, but such methods are not applicable for the design of discrete like micro-robotic appendages. In general, the overall design methodology for the MEMS terrestrial micro-robots is lacking and many of the new design objectives need to be defined.

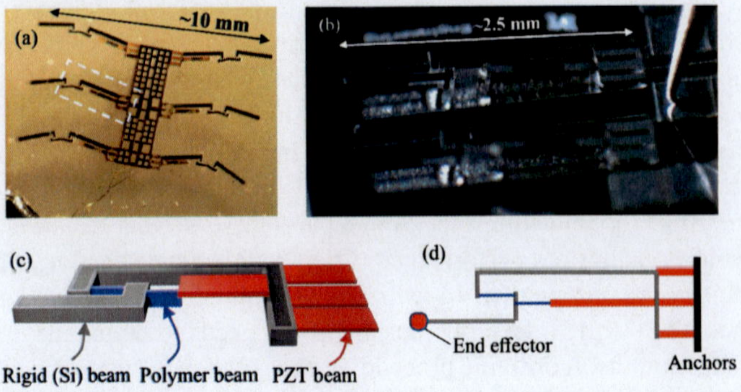

Fig. 1. a) MEMS terrestrial micro-robot prototype based on thin-film piezoelectric actuators with sample appendage design marked; b) micro-robot appendage tests structure composed of PZT; c) illustration of PZT, polymer and silicon structures arranged in sample appendage; d) illustration of appendage model, used in topology optimization framework [7].

In our previous paper [15] the basic topology optimization of the micro-robotic appendages that can realize control of the end-effector in different directions was presented. But only limited design possibilities were explored where the focus was on maximizing control of the appendages end-effector. In this paper a new synthesis method is presented for the realization of micro-robotic appendage designs, considering different performance characteristics like projected walking speed, transport ability, and total power (not done in [15]). The synthesis method will be presented, obtained solutions, together with Finite Element Method characterization of the selected micro-robotic appendage designs.

2 Synthesis of MEMS Terrestrial Micro-robotic Appendages

The main goal of the synthesis is to realize optimal micro-robotic appendage solution with optimal performance properties. In other words, to obtained micro-robotic appendages designs that can realize high walking speeds, have good transport ability, and consume small power. From the synthesis perspective, micro-robotic appendages are comprised of (Fig. 1c and d):

(1) thin-film PZT as active elements, with three layers (top- Au layer, middle- PZT layer, and bottom- SiO_2 layer); when an appropriate voltage is applied the PZT contracts and bends,
(2) Si as rigid passive elements, formed where the silicon substrate is retained,

(3) parylene-C as soft/elastic passive elements; these are formed from combined surface deposition and trench refill, forming beam-like elements with either T- or L- shaped cross-sections.

These elements correspond to the element types that have been co-fabricated with thin-film PZT in [7]. The aim of combining multiple material elements is: Si-rigid elements need to perform substantial weight-bearing, parylene-C soft/elastic element structure needs to amplify the small PZT input strokes, where both soft and rigid element structure should not expose the PZT to excessive forces from robot weight-bearing.

The synthesis of the micro-robotic appendages includes finding the optimal distribution/arrangements (number, type, location, position, orientation) and dimensions (width, thickness, length) of the thin-film PZT, Si and parylene-C so that resulting appendage design can achieve the desired performances (high walking speed and transport ability, with minimal power requirements). To realize such synthesis the topology optimization approach is adopted [10, 13]. Here the designer only needs to define the desired input parameters, then thought algorithmic procedures, the solution is automatically obtained.

The overview of the developed topology optimization of the micro-robotic appendages is illustrated in Fig. 2. The individual synthesis steps include:

(1) defining an initial problem – setting the desired input parameters (Fig. 2a),
(2) parameterization – translation of the defined problem into a set of variables that can be optimized (Fig. 2b and c),
(3) optimization process – applying a search algorithm to find the optimal solution of the defined problem (Fig. 2d, e and f).

In the next subsections, these steps are described in detail.

2.1 Problem Definition

As the first step in the synthesis process, the initial problem needs to be defined (Fig. 2a). This includes setting: the design domain (space in which the appendage solution needs to fit) where here the rectilinear shape is defined, location of the supports (the left end-edge of the designs domain is selected), location of the output (right-top-edge point is selected as end-effector of the appendage), inputs (input voltage is applied to activate the thin-film PZT), external loads that act on the micro-robotic appendage (here external loads are applied in both in-plane and out-of-plane direction), material properties for all the elements- thin-film PZT, parylene-C and Si (Young modulus), range of possible widths of all used active and passive elements, number of thin-film PZT and Si that will be used in the appendage structure, shape of parylene-C (T or L shaped cross-section), together with other constraints and desired objects (like the desired volume of the end solution). The specific parameter values that are used in the synthesis of the micro-robotic appendages are given in Table 1.

2.2 Parameterization

The second step in the synthesis process is to realize the parameterization of the set problem. The design domain needs to be somehow represented with variables that can be optimized. Due to the discrete nature of the micro-robotic appendages design (Fig. 1d), here the ground structure approach (GSA) is adopted for the parameterization. Thus, the design domain is divided by nodes and spanned by beam-like elements connecting these nodes (Fig. 2b). This represents the initial solution within which optimal micro-robotic appendage performance is searched. Initially, all the elements represent passive parylene-C elements, where during optimization some of the elements are replaced with rigid Si elements and some with active thin-film PZT's (Fig. 2c). Elements are simultaneously turned on/off from the initial ground structure (Fig. 2d) until the optimal solution is yielded (Fig. 2e and f). Here multiple variables are defined: a variable that turns on/off paryelene-C elements, variable that defines a selection of active thin-film PZT, variable that defines a selection of passive rigid Si elements, and variable that defines the selection of parylene-C element type (T or L shaped). The width of all the elements is optimized as well, thus one variable is defined for each of the elements that mark a possible range of widths. Additionally, the overall shape of the appendages is optimized by allowing nodes to wander (Fig. 2b) withing the predefined regions (the lengths and position of individual elements are changed). Thus, two variables are introduced that defines the possible range of node wandering in two directions (x and y) in in-plane (the plane of microstructure fabrication on a silicon wafer).

At the beginning, the initial ground structure represents a fully connected structure (Fig. 2b), where all the nodes are interconnected with one beam-like element. Such a ground structure usually leads to solutions that contain overlaying elements (one on top of the other, in two parallel planes). The micro-robotic appendage designs with such elements are difficult to fabricate in the MEMS domain. Thus, a filter is developed that searches and eliminates the overlaying elements from the structure, prior to the optimization. The reduced initial ground structure so obtained contains many overlapping intersecting elements (where elements cross each other while located in two parallel planes, one above the other). Such ground structure often can lead to appendage designs with intersecting elements, which is not desirable. Intersecting/crossing elements usually increase the overall stiffness of the end solution. This means that most of the thin-film PZT input actuation would be spent on deforming the structure thus only a small portion of the input displacement/forces would be transferred to the output. Moreover, appendage designs with intersecting elements are hard to fabricate. To avoid this, a simplified initial ground structure could be used, where one node is connected to only neighboring nodes (meaning that crossing elements do not exist or the number of such elements is small, where at the place of intersection nodes are defined). But such a simple initial net of elements cannot effectivity represent a design domain, and many of the potential good solutions are lost at the beginning. Another solution is to use the reduced initial ground structure but defining the nodes at the intersections, which is not desirable as solutions with small element length could be obtained (difficult to manufacture). Thus, to avoid the overlapping intersecting elements, a filter in the form of an objective goal is developed here. The filter searches and calculates the total number of intersections in the micro-robotic appendage system, where during the optimization this number is minimized.

The variables, initial net size, number of elements, and other parameters used in the synthesis of micro-robotic appendages are given in Table 1.

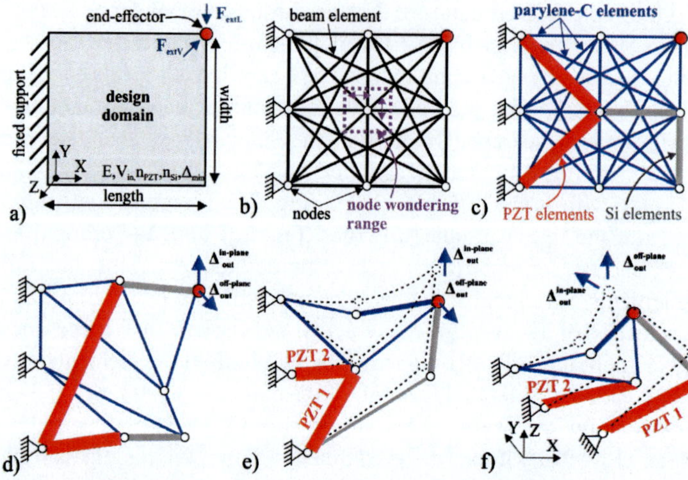

Fig. 2. Synthesis of the micro-robotic appendages: a) problem definition; b) parameterization of the design domain; c) initial ground structure with selected parylene-C, Si and thin-film PZT elements; d) optimization process (selection between different passive and active elements); e) example of the optimal solution shown in an initial (solid lines) and deformed (dash lines) state when all thin-film PZT's are active (planar view); f) optimal solution shown in space in an initial (solid) and deformed state (dashed), where end-effector realize displacement in both in-plane and out-of-plane direction.

2.3 Optimization and Objective Function

As the last step in the synthesis of the micro-robotic appendages, a search algorithm needs to be applied to find the optimal solution of the given problem (Fig. 2e and f). Due to a large and complex design space that contains many variables, an optimization is applied as a search algorithm. Further, for the optimization, an objective function needs to be defined to guide the optimization algorithm towards the desired set goal.

The main aim of the optimization is to realize appendage designs that can achieve relatively high walking speeds, and good transport ability (payload capacity) of the resulting micro-robot while consuming low power. There are fundamental tradeoffs between these performance characteristics, as reducing power consumption reduces both speed and payload capacity, while higher payload capacities generally result in lower speeds. However, there are some opportunities to leverage the differences between internal piezoelectric deformation and point loads at the appendage tip, and dissimilar behavior in different axes to improve system-level performance. Thus, in the optimization, the walking speed and transport ability need to be maximized while the total required power need to be minimized. The resulting objective function for the synthesis of micro-robotic appendages is given in a form:

$$\max[T_{\text{ran}} \cdot w_1 + V_{\text{elW}} \cdot w_2 - P_{\max} \cdot w_3 - d_{\text{ext}} \cdot w_4 - n_{\text{int}} \cdot w_5] \tag{1}$$

The Eq. (1) is formulated so that transport ability (T_{ran}) and projected walking speeds (V_{elW}) are maximized, and total need power ($P_{\max}$) is minimized. Further to increase the appendage payload capacity the end-effector displacement due to external loads (d_{ext}) is minimized, and total number of intersections (n_{int}) are minimized as well. The individual terms in the Eq. (1) are calculated by following set of equations:

$$T_{ran} = \left(\frac{F_{extV} \cdot d_{elV}}{d_{FextV} \cdot 9.8} - \frac{\sqrt{\frac{F_{exL} \cdot d_{FextV} \cdot 9.8}{d_{FextL} \cdot F_{extV} \cdot d_{elV}}} \cdot \pi \cdot C_{ap} \cdot V_{pzt}^2}{2 \cdot B_{pw}} \right) \cdot V_{elW} \tag{2}$$

$$V_{elW} = \sqrt{\frac{F_{exL} \cdot d_{FextV} \cdot 9.8}{d_{FextL} \cdot F_{extV} \cdot d_{elV}}} \cdot \pi \cdot d_{elL} \Big/ 2 \tag{3}$$

$$P_{max} = n_{leg} \cdot \sqrt{\frac{F_{exL} \cdot d_{FextV} \cdot 9.8}{d_{FextL} \cdot F_{extV} \cdot d_{elV}}} \cdot \pi \cdot C_{ap} \cdot V_{pzt}^2 \Big/ 2 \tag{4}$$

In brief Eq. (2) is based on the maximum deflection permitted due to gravity relative to appendage vertical range of motion, Eq. (3) is based on estimated natural frequency and appendage lateral range of motion, and Eq. (4) is based on estimated natural frequency and total PZT capacitance, where:

d_{elV} – end-effector displacement in the vertical direction (out-of-plane),

d_{elL} – end-effector displacement in the lateral direction (in-plane),

d_{FextV} – end-effector displacement in vertical direction (out-of-plane) due to external loads,

d_{FextL} – end-effector displacement in the lateral direction (in-plane) due to external loads,

V_{pzt} – input voltage for the thin-film PZT's (Table 1),

B_{pw} – assumed battery power density (Table 1),

C_{ap} – capacitance (calculated by the standard equation that can be found in [16]),

n_{leg} – number of appendages.

All the terms in the objective function are calculated by using linear analysis which is implemented in the optimization algorithm. It is important to mention that this synthesis of micro-robotic appendages represents a multiobjective optimization problem, which is reduced to a single-objective problem by using weighting constants (w_i).

Among many different possible optimization methods, here a Genetic Algorithms (GA) are adopted [9, 13]. GA's are very efficient in finding a global optimum over a large design space that contains many variables. Further, GA's represent a discrete optimization method, thus they are very suitable for optimization of discrete-like appendage designs where the variables are discrete as well. Moreover, a GA's are extensively used in the topology optimization problems [9, 13]. The genetic parameters that are used for the synthesis of the micro-robotic appendages are an initial population of 200, a total of 1000 generations, selection function type roulette, crossover probability of 95%, elite count of 2 members, and mutation probability of 9%.

2.4 Solutions of Micro-robotic Appendages

Multiple GA's optimizations are run for a different set of initial parameters and values of weighting constants (Table 1) to obtain the micro-robotic appendage solutions. The

Table 1. Parameters used for the synthesis of the micro-robotic appendages.

Parameter	Case 1	Case 2
Design domain	2000 μm × 2000 μm	
Grid size	3 × 3	
Number of parylene-C el.	28	
Number of thin-film PZT and Si el.	2	
Input voltage for thin-film PZT	15 V	
Element modulus	parylene-C, $E_{pary} = 1.3$ GPa; Si, $E_{Si} = 170$ GPa; PZT, $E_{base} = 80$GPa, $E_{PZT} = 56$ GPa, $E_{top} = 60$ GPa	
Element out-of-plane thickness	$t_{pary\text{-}T} = 70$ μm; $t_{pary\text{-}L} = 5$ μm; $t_{Si} = 30$ μm; $t_{base} = 0.5$ μm, $t_{PZT} = 1$ μm; $t_{top} = 1$ μm	
Range of element widths	$w_{pary\text{-}T} = 3, 5, 7, 9$ μm; $w_{pary\text{-}L} = 20, 60, 100$ μm $w_{Si} = 10, 40$ μm $w_{PZT} = 50, 100, 150$ μm	
Node wandering space	50 × 50 μm	100 × 100 μm
External loads	$F_y = 1$ μN, $F_z = 1$ μN	
Thin-film PZT el. Range	PZT1 = 1 - 28 PZT2 = 1 - 28	
Si el. Range	Si1 = 1 - 28 Si2 = 1 - 28	
Parylene-C el. Type	parylene-T type = 1 - 28 parylene-L type = 1 - 28	
Assumed battery power density	$B_{pw} = 100$ (W/kg)	
Weighting constants	w_1 = 1e9 - 5e9, w_2 = 1e3 - 2e3, w_3 = 0.3e5- 4e5, w_4 = 0.8, w_5 = 1e3	

optimization process starts with initial ground structure (Fig. 2b), where simultaneously some parylene-C elements are turned on/off from the structure, some are chosen as active thin-film PZT and some as passive rigid- Si elements, the width of individual elements are selected together with node locations/positions (Fig. 2c and d). The remaining parylene-C, thin-film PZT, and Si elements with corresponding dimensions and node locations form the optimal topology of the micro-robotic appendages (Fig. 2e and f).

Figure 3 shows some of the obtained solutions of the micro-robotic appendages. In all cases, the appendages realize appropriate output displacement of the end-effector, when all thin-film PZT's are activated at once. The performance characteristic for each obtained solution is depicted in Fig. 4. The appendage designs are compared against realized walking speed and required powered when end-effector displacement is considered. The results show that optimal solutions are very sensitive to weighting of the various performance metrics, but we see a tendency for locally optimal solutions to fall into three major categories: high stiffness, low displacement solutions with high transport capacity and high natural frequency, keeping speed high but with typically small step lengths and high power consumption (i.e. appendage 1 in Fig. 3, and solutions elevated on the z-axis of Fig. 4); low stiffness, high displacement solutions with modest projected walking speeds and large range-of-motion, but low transport capacity (i.e. appendages 2, 4 in Fig. 3, clustered near lower right of Fig. 4); and low displacement solutions having reduced active element size and natural frequency, for extremely low-power but low-speed locomotion (i.e. appendage 3 in Fig. 3, clustered near lower left of Fig. 4).

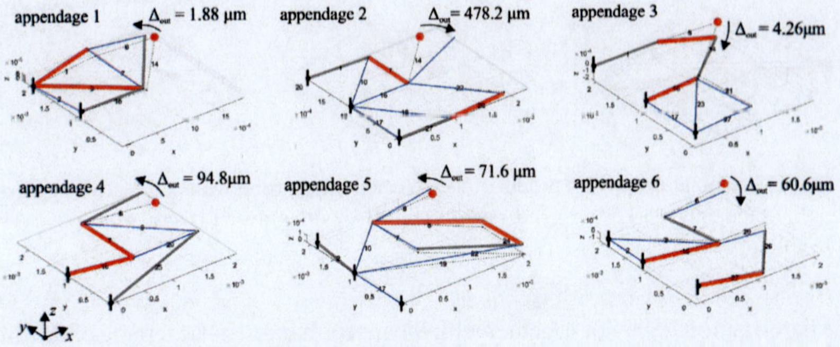

Fig. 3. Topology optimized solutions of the MEMS terrestrial micro-robotic appendages (a few representative obtained designs are shown).

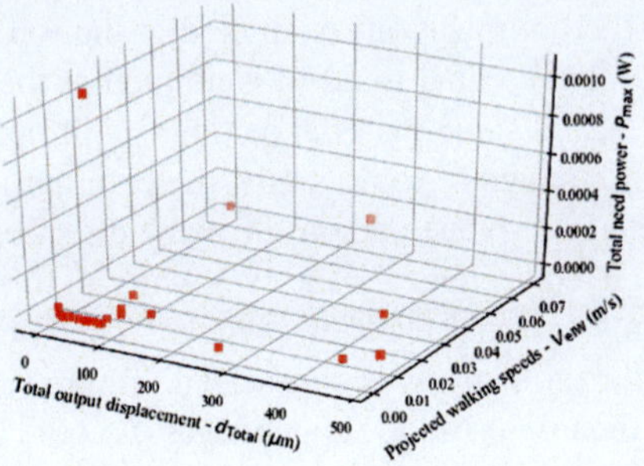

Fig. 4. The performance characteristics of the micro-robotic appendage solutions obtained with topology optimization.

3 FEM Based Analysis of Micro-robotic Appendages

To verify the micro-robotic appendage solutions (Fig. 3) Finite Element Method simulations are performed by using *COMSOL Multiphysics*. Based on the obtained solutions with optimization (Fig. 3), a 3D solid model of the appendages is designed considering the fabrication requirements. Here only a few of the selected designs are presented (Fig. 5). Like Fig. 1 and Fig. 3, micro-robotic appendages are comprised of:

- thin-film PZT active elements with three layers (top- Au, middle- PZT, bottom- SiO_2); marked in red (Fig. 5),
- Si passive rigid elements, enclosed in SiO_2 trenches; marked in gray (Fig. 5),
- parylene-C passive soft elements; marked in blue (Fig. 5).

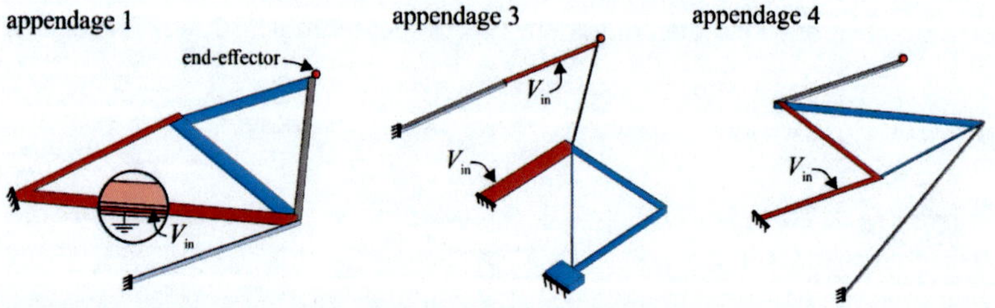

Fig. 5. A FEM setup and solid 3D model of the selected micro-robotic appendage designs, active thin-film PZT elements marked with red, passive (rigid) Si with gray, and passive (soft) parylene-C with blue color.

The deformation behavior together with the performance characteristic of the micro-robotic appendage designs is investigated. For the simulations in *COMSOL*, the stationary structural mechanics coupled with the piezoelectric module are used. The FEM setup is as follows: the fixed boundary condition (Fig. 5) is applied at the same location as in the corresponding topology optimized solution (Fig. 3); an input electric voltage of 15 V (the same value used in the optimization) is applied at the top surface of the middle-PZT layer, and ground (voltage equal to zero) is applied at the bottom surface of the middle-PZT layer (Fig. 5); the same material settings as in optimization are used for the corresponding type of elements. Additionally, the external loads are applied at the end-effector location, both in in-plane and out-of-plane direction. Two simulations are run for each of the investigated designs: for the case when all thin-film PZT's are active, and for the case when only external loads are applied. The end-effector displacement is read out while the nonlinear analysis is considered. Figure 6 shows the result of the deformation behavior of the micro-robotic appendages. It could be concluded that in all the simulated cases, the appendages realize appropriate output displacement of the end-effector when the input voltage is applied i.e. thin-film PZT's are activated. The results further show that similar deformation behavior is realized as in the case of topology optimization solutions (Fig. 3). The appendage performances are depicted in Fig. 7, where designs are compared against realized walking speed (V_{elW}) and needed total power ($P_{\max}$) considering total output displacement of the end-effector (d_{Total}). The results in Fig. 7, are calculated based on the read-out values of the end-effector displacement and using Eq. (2), (3) and (4). The results further demonstrate representative designs of the types described in Sect. 2.4, with closest agreement between the low-power, low-speed appendage 3. The larger-displacement designs, however, generate substantially less motion in Comsol simulation. We attribute this primarily to the difference between idealized connections between beam elements in topology and distributed intersections present in the solid model, with greater resistance to torsional motions and to lateral bending motions present between wider beams when fully modeled by FEA. Notably, projected walking speed does not decrease, because the reduced step length is offset by increased natural frequency (increased feasible step frequency) due to the increased stiffness in FEA results.

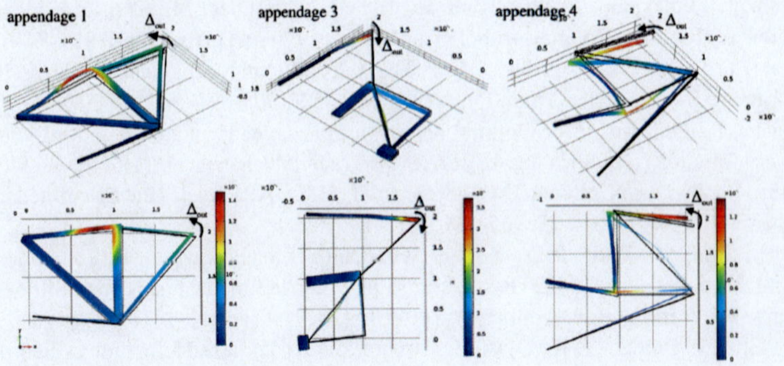

Fig. 6. FEM simulation results for the deformation behavior of the micro-robotic appendage solutions when the input voltage of 15 V is applied.

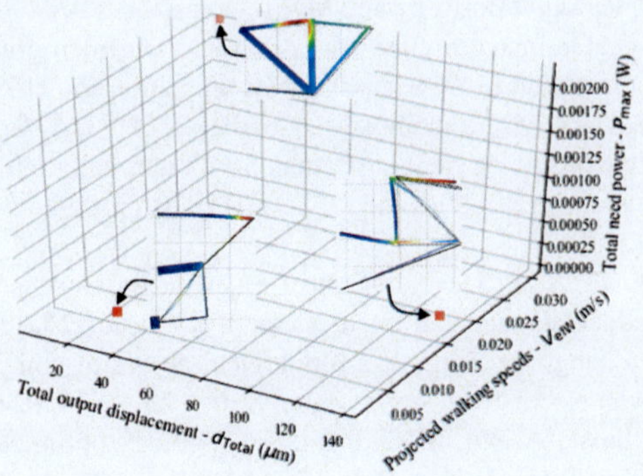

Fig. 7. FEM simulation results for performance characteristic of the selected micro-robotic appendage solutions.

4 Conclusions

This paper presents a novel synthesis method for the MEMS micro-robots and their corresponding appendages. It was shown that by using the developed optimization method a variety of solutions for appendage design could be realized in response to varying weighting of candidate performance specifications, such as transport ability, walking speed, range-of-motion, and/or power budget. Future work may analyze combined behavior through metrics such as cost-of-transport, detailed gait analysis and control in simulation of full multi-legged robots, and physical prototyping of select appendage and robot designs.

References

1. Palagi, S., Fischer, P.: Bioinspired microrobots. Nat. Rev. Mater **3**, 113–124 (2018)
2. Sitti, M., Wiersma, D.S.: Pros and cons: magnetic versus optical microrobots. Adv. Mater. **2020**, 1906766 (2020)
3. Oldham, K., Pulskamp, J., Polcawich, R., Ranade, P., Dubey, M.: Thin-film piezoelectric actuators for bio-inspired micro-robotic applications. Integr. Ferroelectr. **1**(95), 54–65 (2007)
4. Huber, J.E., Fleck, N.A., Ashby, M.F.: The selection of mechanical actuators based on performance indices. Proc. Math. Phys. Eng. Sci. **453**(1965), 2185–2205 (1997)
5. Edamana, B., Oldham, K.R.: Optimal low-power piezoelectric actuator control with charge recovery for a microrobotics leg. IEEE/ASME Trans. Mechatron. **18**(1), 251–262 (2013)
6. Murthy, R., Das, A.N., Popa, D.O., Stephanou, H.E.: ARRIpede: an assembled die-scale microcrawler. Adv. Robot. **25**, 965–990 (2011)
7. Choi, J., Shin, M., Rudy, R.Q., Kao, C., Pulskamp, J.S., Polcawich, R.G., Oldham, K.R.: Thin-film piezoelectric and highaspect ratio polymer leg mechanisms for millimeter-scale robotics. Int. J. Intell. Rob. Appl. **1**(2), 180–194 (2017)
8. Qu, J., Choi, J., Oldham, K.R.: Dynamics structural and contact modeling for a silicon hexapod microrobot. J. Mech. Rob. **9**(6), 061006 (2017)
9. Howell, L.L., Magleby, S.P., Olsen, B.M.: Handbook of Compliant Mechanisms. Wiley, Chichester (2013)
10. Saxena, A.: Topology design of large displacement compliant mechanisms with multiple materials and multiple output ports. Struct. Multi. Optim. **30**, 477–490 (2005)
11. Ruiz, D., Sigmund, O.: Optimal design of robust piezoelectric microgrippers undergoing large displacements. Structural and Multidisciplinary Optimization **57**, 71–82 (2018)
12. Moussa, R.E.K., Grossard, M., Boukallel, M., Chaillet, N., Hubert, A.: Observation-oriented design of a monolithic piezoelectric microactuator with optimally integrated sensor. In: 41st International Symposium on Robotics and ROTOTIK 2010, pp. 954–961. Verlag 1, June 2010, Munich, Germany (2010).
13. Milojević, A., Pavlović, N.D.: Development of a new adaptive shape morphing compliant structure with embedded actuators. J. Intell. Mater. Syst. Struct. **27**, 1306–1328 (2016)
14. Wang, Y., Luo, Z., Zhang, X., Kang, Z.: Topological design of compliant smart structures with embedded movable actuators. Smart Mater. Struct. **23**(4), 045024 (2014)
15. Milojević, A., Krokhmal, V., Wu, B., Oldham, K.: Topology optimization of Micro-robotic appendages combining Piezoelectric, Polymer, and Silicon Beams. In: 2019 International Conference on Manipulation, Automation and Robotics at Small Scales (MARSS), pp. 1–6. Helsinki, Finland (2019)
16. Oldham, K.R., Pulskamp, J.S., Polcawich, R.G., Dubey, M.: Thinfilm PZT lateral actuators with extended stroke. J. Microelectromech. Syst. **17**(4), 890–899 (2008)

Author Index

L. Zentner and S. Strehle (Eds.): MAMM 2020, MMS 96, p. 149, 2021.
https://doi.org/10.1007/978-3-030-61652-6